Bacterial Systematics

Bacterial Systematics

NIALL A. LOGAN BSc PhD
Department of Biological Sciences
Glasgow Caledonian University

OXFORD
Blackwell Scientific Publications
LONDON EDINBURGH BOSTON
MELBOURNE PARIS BERLIN VIENNA

Editorial Offices:
Osney Mead, Oxford OX2 0EL
25 John Street, London WC1N 2BL
23 Ainslie Place, Edinburgh EH3 6AJ
238 Main Street, Cambridge
Massachusetts 02142, USA
54 University Street, Carlton
Victoria 3053, Australia

Other Editorial Offices:
Librairie Arnette SA
1, rue de Lille
75007 Paris
France

Blackwell Wissenschafts-Verlag GmbH
Düsseldorfer Str. 38
D-10707 Berlin
Germany

Blackwell MZV
Feldgasse 13
A-1238 Wien
Austria

First published 1994

Set by D & N Publishing, Ramsbury, Wilts
Printed and bound in Great Britain
at The Alden Press, Oxford

DISTRIBUTORS

Marston Book Services Ltd
PO Box 87
Oxford OX2 0DT
(*Orders*: Tel: 0865 791155
Fax: 0865 791927
Telex: 837515)

USA
Blackwell Scientific Publications, Inc.
238 Main Street,
Cambridge, MA 02142
(*Orders*: Tel: 800 759-6102
617 876-7000)

Canada
Oxford University Press
70 Wynford Drive
Don Mills
Ontario M3C 1J9
(*Orders*: Tel: 416 441-2941)

Australia
Blackwell Scientific Publications Pty Ltd
54 University Street
Carlton, Victoria 3053
(*Orders*: Tel: 03 347-5552)

A catalogue record for this title
is available from the British Library

ISBN 0-632-03775-X

Library of Congress
Cataloging in Publication Data

Logan, Niall A.
Bacterial systematics/Niall A. Logan.
p. cm.
Includes bibliographical references and index.
ISBN 0-632-03775-X
1. Bacteriology—Classification. I. Title.
QR81.L64 1994
589.9'0012—dc20

Contents

Preface

I have long felt the need for an introductory text covering the theory and practice of bacterial classification and identification, and this is my attempt at such a book. It was written at a time of rapid progress in the subject, and while the use of molecular sequencing techniques in evolutionary studies has tended to dominate the literature, it has also revived interest in all the other aspects of bacterial systematics.

I thank the authors and publishers who kindly allowed me to adapt their figures for this book, and am most grateful to the following people who generously provided material for illustrations: Dr G.K. Adak, PHLS Communicable Disease Surveillance Centre (Fig. 8.2); Mr J. Bagshaw, bioMérieux UK (Figs 5.3a & 5.5); Dr R.C.W. Berkeley, University of Bristol (Fig. 2.8); Dr P. Berry, Mercia Diagnostics (Fig. 5.3c); Mr N. Claxton, Becton Dickinson, UK (Fig. 5.3b); Dr P. De Vos, Universiteit Gent (Figs 2.4, 2.5 & 3.6); Dr M. Gillis, Universiteit Gent (Fig. 6.3); Prof. M. Goodfellow, University of Newcastle upon Tyne (Fig. 14.2); Dr A.C. Hill , Medical Research Council Toxicology Unit (Fig. 13.1); and Dr C.W. Moss, Centers For Disease Control, Atlanta (Figs 2.2 & 9.8); I also thank Mr J. McWilliams, Glasgow Caledonian University, for the photography. I am indebted to Prof. P.H.A. Sneath, University of Leicester, for explaining to me the taxon-radius model for identification, to Prof. M. Goodfellow for reading the manuscript and offering much helpful comment, and to my wife and children for their forbearance.

NIALL A. LOGAN

1 Introduction

Not Chaos-like, together crush'd and bruis'd,
But as the World, harmoniously confus'd:
Where Order in Variety we see,
And where, tho' all things differ, all agree.

Alexander Pope, 1713, *Windsor Forest*

The development of bacterial classification

Michel Adanson's proposal in 1764 for natural classifications based upon many characters and his contempt for 'systems' based upon a selected few brought him into conflict with Carl von Linné (Linnaeus), who had devised such systems for plants, animals and minerals. Linnaeus doubted the value of microscopy and was therefore unable to classify the animalcules described by Antonie van Leeuwenhoek and others for want of characters; he placed them all in a class of invertebrates which he called 'Chaos' — the shape of matter before it was reduced to order.

Most microscopists of the 17th and 18th centuries did not try to classify the infusion animalcules that they observed and often described so meticulously. With works published in 1773 and 1774 the Danish naturalist Otto Müller was the first to attempt a systematic arrangement of the animalcules, but he did not make a clear distinction between what we now call protozoa and bacteria. For his last work however, which was published posthumously in 1786, he did create two form genera, *Monas* and *Vibrio*, which contained bacteria and accommodated the punctiform and elongated types.

In 1838 Christian Ehrenberg considerably extended Müller's nomenclature and added the helical bacteria. He was handicapped by the microscopes of his day, and many of the groups that he described cannot now be recognized. None the less, he established genus and species names, such as *Spirochaeta plicatilis* and *Spirillum volutans*, that are still in use. Subsequent workers devised simpler classifications, but all these early bacterial systematists based their arrangements upon microscopic morphology, and assumed constancy of form at a time when the theories of spontaneous generation and pleomorphism (i.e. no constancy of form) were still widely held and the germ theory of disease had yet to be proven.

In the 1870s Ferdinand Cohn, having satisfied himself that bacterial forms were constant irrespective of environmental conditions, recognized the existence of a wide diversity of bacteria but considered them to form a distinct group that was unrelated to fungi, yet had close affinities with blue-green algae. He arranged bacteria in six form genera, which he believed to be the natural ones, and many provisional species, but appreciated that the physiologies, products and pathogenicities of similar-shaped organisms might differ, and so used such properties in subdividing his genera. In *Bacillus* he placed *B. subtilis*, having recognized its spores as persistent forms, and another sporeformer, *B. anthracis*, whose life history Robert Koch demonstrated to him in 1876 — thereby proving the germ theory of disease. From his studies of infectious diseases, Koch later concluded that the different forms of pathogenic bacteria must be regarded as distinct and constant species.

Cohn found it necessary to defend his beliefs, based as they were on careful observation, against those who still tried to prove, using defective methods, that bacteria were merely stages in the development of fungi (after the foundation of medical mycology in the 1840s several diseases, including cholera and measles, had been attributed to fungi by some workers), and that changes in environmental conditions altered bacterial morphology. In 1882 Edouard Buchner claimed to have converted *B. subtilis* to *B. anthracis* by shaking it in media at different temperatures!

Pure cultures

The concept of bacterial species led to the idea that pure cultures might be obtainable. By 1872 Cohn's co-worker Joseph Schroeter had cultivated pure colonies of chromogenic bacteria, including the violet-pigmented organisms now informally referred to as chromobacteria, on a variety of starchy foods, eggs and meat, and in 1878 Joseph Lister obtained a pure culture of a milk-souring organism by dilution. Koch had turned from using animal passage for purifying his pathogenic strains to developing plates of cultivation media solidified by gelatin, which was subsequently replaced by agar, and he demonstrated and published his methods in the early 1880s. So began the 'golden age of microbiology' — bacteria could now be isolated routinely by streak dilution culture and, as Perkins observed in 1928, this 'discovery of the principles of pure-culture study resulted in such a sudden burst of investigation that it was a lost month in which a new organism was not described, catalogued, and laid away, very frequently in the wrong grave', and led to the development of the many diverse characterization tests, such as the Voges–Proskauer (1898) test for acetylmethylcarbinol, the methyl-red test (1915) for production of copious acid from glucose, and the tests for cytochrome oxidase (1928) and urea hydrolysis (1946), that were described over the next 80 years.

The various later workers who developed classifications drawn up by Cohn continued to regard spherical organisms (cocci) as primitive forms and they emphasized morphological characters, especially shape and size of cells, arrangement or absence of flagella, and production of spores, in their higher divisions of bacteria. This Linnaean approach caused confusion; *B. subtilis* and *B. anthracis*, for example, were put together as sporeformers, or simply as rods by some, and placed in separate genera by others because the former is motile and the latter is not. It was only in the second decade of the twentieth century, following the work of Orla-Jensen, and with the reports of the Society of American Bacteriologists' Committee on Bacterial Classification and Nomenclature in 1917 and 1920, that physiological characters such as aerobic or anaerobic growth became widely used in descriptions of genera.

Bergey's Manual

Bergey's Manual of Determinative Bacteriology, published in 1923, was written to provide a modern identification key for bacteria but little of it was based on direct experience of the organisms and this first and five subsequent editions, which were very difficult to use in the laboratory unless organisms were already identified to genus level, came to be used more as the authoritative reference works on bacterial classification. In successive editions of *Bergey's Manual*, and in the absence of a usable fossil record, a quasi-evolutionary approach to arrangement, as used for plants and animals, was adopted; in the seventh edition, published in 1957, the photoautotrophs were regarded as the most primitive forms and the rickettsias, to which the viruses were tentatively attached, the most advanced. In the eighth edition, which was published in 1974, it was considered that such an approach was no longer justifiable and the genera, sometimes grouped in families where thought useful, were arranged in 19 parts. Each of these was based upon a few readily determined characters and bore a vernacular name such as 'Gram-positive cocci'; hence, no evolutionary relationships were implied. Instead, a tentative subdivision of the Kingdom *Procaryotae* was given — proposals for the recognition of such a kingdom of anucleate organisms had been made since the late 1930s and were subsequently supported by microscopy, including electron microscopy, and molecular studies.

Numerical taxonomy

As the variety of methods for characterizing bacteria increased, so bacterial systematists suffered more and more from the lack of quantitative approaches to classification. In 1957 Peter Sneath revolutionized the subject with his two papers 'Some Thoughts on Bacterial Classification' and 'The Application of Computers to Taxonomy' (**taxonomy** being the science of classification; it is often used as a synonym for classification) in which he described numerical methods of grouping bacteria using the chromobacteria together with some other Gram-negative rods as his example.

In subsequent publications, including two books (1963, 1973) written in collaboration with Robert Sokal, Sneath summarized the fundamental position of **numerical taxonomy** (the grouping by numerical methods of taxonomic units based upon their character states), in principles referred to as neo-Adansonian and which included the need for many and equally-weighted characters. The rapid development and availability of computers enabled the integration of many different pieces and types of data into the classification process, which consequently benefited from greater information content and the more objective recognition of groups or **taxa**; thus, the realization of Adanson's ideas after 200 years depended, to a great extent, on the advent of electronic computing.

Modern methods

The period that saw the development of numerical taxonomy also saw the rise of chemotaxonomy – the application of modern biochemical analytical techniques, principally chromatographic and electrophoretic separation methods, to the study of distributions of specific chemicals such as amino acids, proteins, sugars and lipids in bacteria. Of particular interest are studies of nucleic acids, the ultimate objective being rapid and direct sequencing. Data from DNA–rRNA reassociation experiments and protein sequencing, and from methods of inferring and comparing sequences of rRNA molecules, which have evolved very slowly so that the base sequences of many cistrons are highly conserved, have been used like a bacterial fossil record (but without time units) to facilitate the construction of prokaryote '**phylogenies**' (genealogical trees) that are quite different to those implied in the seventh edition of *Bergey's Manual*.

The first of four volumes of *Bergey's Manual of Systematic Bacteriology*, published in 1984, was arranged in sections much like the parts of the eighth edition of *Bergey's Manual of Determinative Bacteriology*, and proposed a revision of the higher taxa of prokaryotes consistent with the phylogenetic information then available. It is becoming clear that most existing bacterial classifications, which are based upon **phenotypic characters** (observable expressions of the **genotype**) with a view to providing identification schemes, correlate poorly with the evolutionary relationships that appear to exist between the higher taxa. The problem now facing taxonomists is the construction of a single and practical scheme capable of incorporating genotypic and phenotypic information. Consequently, **polyphasic** taxonomic studies, which use ranges of both genomic and phenotypic approaches, are now widely favoured.

Why classify bacteria?

Classification, the grouping of things together, is a common and important human activity that has been practised since earliest times. In dealing with large numbers of objects or pieces of information some convenient system of orderly arrangement is needed for the purposes of storage and retrieval, and in any scientific work up-to-date classifications are essential. Books, for example, may be grouped by author, subject, title or a combination of these; without such a system a library would be virtually unusable and the retrieval of information from it hopelessly inefficient.

In bacteriology, classification is a means of summarizing our knowledge of prokaryotes and cataloguing that knowledge. As this information is constantly and rapidly expanding, so classifications evolve and increase in importance, with contemporary schemes reflecting our state of knowledge about the organisms concerned. Such schemes may themselves be classified into **natural** and **artificial** (or **special purpose**) types.

Natural classifications

What constitutes a natural or logical classification is hard to define for any group of organisms; as the late Sam Cowan commented in 1978: 'the biologist . . . is attuned to the vagaries of living things and does not expect an experiment to be exactly repeatable . . . is surprised by the expected, and astounded by the fulfilment of a forecast or prediction. Thus, if we say that a classification is logical we know that it is not a classification of biological units, whereas if we say that it is expedient we will be ready to lay odds on its being the work of a biologist dealing with units of his (or her) own creation' — the bacteriologist must never forget that genera and species are artificial concepts and that the bacteria show no interest in their classification.

In the pre-Darwin era Aristotle's theories were applied, the idea being to classify according to an item's 'essential nature' by repeated subdivision, but after the publication of Darwin's theories an evolutionary approach, using fossil records, was generally adopted. However, the application of this **phyletic** method to bacteria was impractical, not only because of the absence of usable fossil evidence, lack of information about the extent of convergent evolution, and the simplicity of bacterial morphology, but also because of the breadth of biological variation that occurs in such rapidly reproducing organisms owing to mutation, adaptability and lateral gene transfer. Until recently, therefore, most 'natural' classifications of bacteria were **phenetic** (from *phen*otype and gen*etic*); they were based upon studies of large numbers of properties analysed by numerical methods, and did not attempt to imply evolutionary relationships. The groups recognized by these methods are described as **polythetic**; they are formed by their members having large numbers of characters in common, and they tolerate limited numbers of the exceptional characters that might be expected as a result of biological variation.

Artificial classifications

The term 'artificial' was formerly applied to those classifications that were made for single, clearly defined purposes such as distinguishing between the pathogenic member of a genus and the rest and they could be **monothetic**, that is to say based upon a single defining character. Now that phyletic classifications of bacteria are being attempted, as a result of the molecular studies referred to earlier, and because of advances in identification methods and increasing awareness of the dangers of drawing rigid boundaries between, for example, pathogens and non-pathogens, the definitions and applications of the different types of microbial classifications are changing.

We see, therefore, that even classifications of classifications have to evolve. Natural or phylogenetic classifications of bacteria are now based upon genotypic information, and phenetic taxonomies might be regarded as artificial and **general-purpose** when polythetic but **special-purpose** when monothetic. Until a workable and useful unified classification is constructed, we must consider the aims of phylogenetic and phenetic schemes separately.

Purposes of classifications

Phylogenetic Schemes. Because of the problems outlined above, notions of bacterial evolution have played little part in the development of microbiology or of general evolutionary theory and yet the evolutionary range and history of bacteria greatly exceeds that of eukaryotes. The evolution of bacteria over at least 3.5 billion years spans most of the earth's 4.6 billion year history (Fig. 1.1) and occurred in step with its geochemical development. Also, the endosymbiont theory suggests that bacteria played an important part in the evolution of eukaryotic cells, with chloroplasts and mitochondria having prokaryote ancestors. Thus, although we lack a time scale, the study of bacterial phylogeny not only helps us to establish a natural bacterial classification and to understand the evolution of bacteria, but also offers important insights into fundamental aspects of biology in general — the origin of cellular life and the development of eukaryotes.

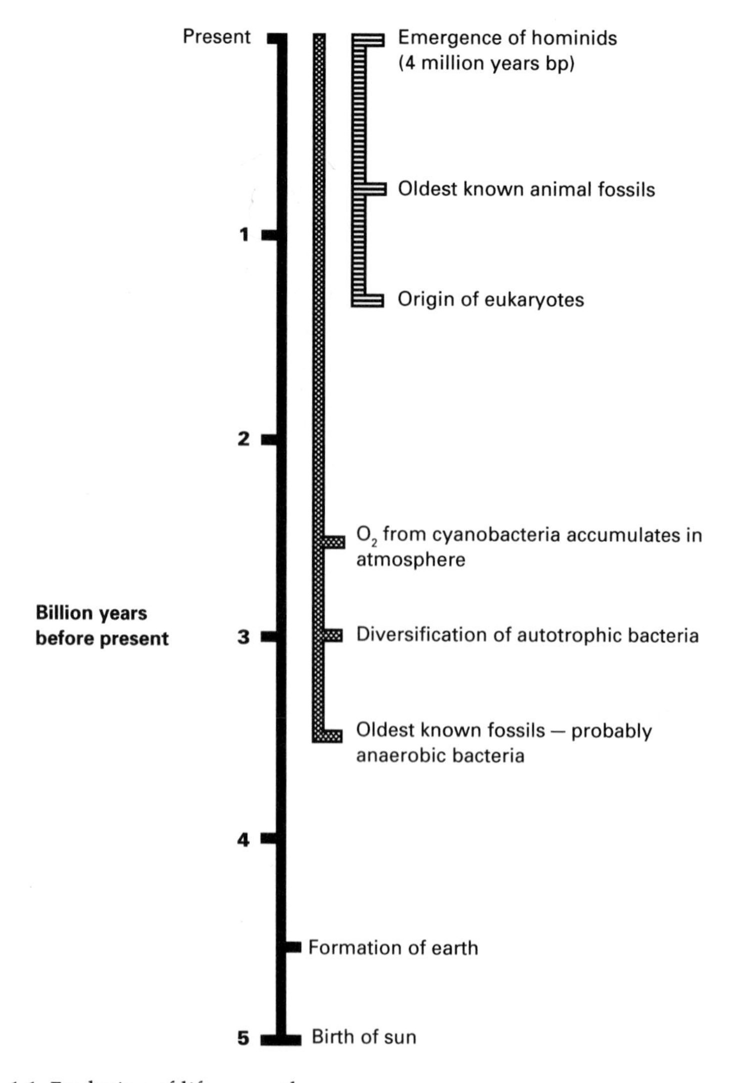

 Fig. 1.1 Evolution of life on earth.

Phenetic Schemes. These classifications have two closely related purposes – identification and prediction. The first of these is a narrowing-down exercise, working towards a name for the unknown organism by process of elimination; having done this, we can move outwards by exploring the available information on other strains of that species and so make general predictions about its properties.

If a classification accurately reflects the overall similarities of its constituent members it will lead to dependable identifications and be efficient in data storage and retrieval; it will summarize the properties of the organisms well and so have a high predictive value. An identification system can be no better than the classification from which it is derived and a good general-purpose classification will have been based upon a large amount of diverse information. With a poor classification based upon few criteria, therefore, identification may be slow and difficult and even if it happens to be successful in putting a name to the organism, the meagre information obtained might be of little value.

When looking at a bacterial cell through a microscope we learn little about the nature of the organism (as the early bacteriologists found) but if, say, from its source, morphology and staining reaction it can be assigned to a group already so clearly defined and described that we know there are no other organisms with which it might be confused, then we can make useful predictions about it.

Imagine, for example, a medical bacteriologist finding Gram-negative diplococci in a specimen of urethral pus, or acid-fast rods in sputum; he or she can immediately suggest respective diagnoses of gonorrhoea and tuberculosis and treatment of the patients can begin without delay. In such cases satisfactory classifications have facilitated rapid and reliable identifications of *Neisseria gonorrhoeae* and *Mycobacterium tuberculosis* on the basis of a few characters, and their significances, the initial treatment regimens, and the appropriate methods of confirming the diagnoses and of establishing antibiotic sensitivities have been predicted by reference to our considerable knowledge of these organisms.

Other qualities

So far the importance of having a high information content has been stressed, but to be effective a classification must also satisfy several other conditions. The first of these is stability, because if the members and definitions of the various taxa keep changing then the classification will be of little practical value and only serve to confuse the user. This stability cannot be absolute, however, the second condition being that as new knowledge is acquired it must be reflected by progress in classification. Satisfying both of these conditions poses problems, but good schemes will often absorb new information and members without the need for radical changes. Last, the classification should of course be founded upon reproducible scientific experiments and so have an empirical rather than theoretical basis; this makes for slow but sound progress.

Nomenclature — 'the handmaid of taxonomy'

We have seen how the process of identification leads to a name which then grants access to further information about the organism. A name thus acts as a means of communication, and the speed and clarity with which it transmits the intended information are the criteria by which a system of nomenclature is judged; a name must therefore be unambiguous, universal and stable. It need not, however, be descriptive because it is merely a label — a point made more eloquently by Shakespeare: 'What's in a name? that which we call a rose / By any other name would smell as sweet'.

An organism may have many different common names around the world and even within a single language there may be several according to region or the background of the person referring to it; furthermore, the same name may be loosely applied to different organisms. Such vernacular names are therefore unsuitable for biological purposes and scientific names, regulated by **codes of nomenclature**, are used instead. It must be appreciated that there is no single 'official' or 'correct' classification of living things and so more than one name can exist for an organism; furthermore, progress in classification often makes name changes necessary. The codes aim to minimize the confusion that may occur as a result.

The species concept and lower ranks

Current systems of biological classification and nomenclature have the **species** as their basic units but this is an artificial concept, it does not exist as an entity, and, especially in bacteriology, it is without a satisfactory definition at present. A bacterial species may be viewed as a group of **strains** (clones derived from single isolations in pure culture) having many features in common and differing considerably from other strains. Interpretation of such a definition is very subjective and depends on the judgment (and prejudice) of the individual investigator.

Taxonomists are often classified as 'lumpers', who emphasize the similarities between their strains and wish to keep names to a minimum, and 'splitters' who stress differences and believe that the clarity inherent in small groups is more important. Bacteriologists may tend to be splitters of the groups they know most about and lumpers of everything else. Rather than being based on high phenetic similarities (yielding **taxospecies**), a bacterial species could be better defined on the basis of genetic relatedness, in terms of DNA relatedness (**genomic species**), or interbreeding, as shown by genetic transfer (**genospecies**) for example, but the application of such definitions might be at variance with the stabilities of many well-accepted groups. Cowan summarized thus: 'a species is a group of organisms defined more or less subjectively by the criteria chosen by the taxonomist to show to best advantage and as far as possible put into practice his (or her) individual concept of what a species is'.

Species may be divided into subspecies on the basis of consistent phenotypic variations or genetically defined groups of strains; *Treponema pallidum* for example, is split into the three subspecies *T. pallidum* subsp. *pallidum, T. pallidum* subsp. *endemicum* and *T. pallidum* subsp. *pertenue*, because although they show 100% DNA relatedness with each other they cause three quite different human diseases, venereal (and congenital) syphilis, endemic syphilis, and yaws respectively, and have different infectivities for laboratory animals.

Categories below subspecies are often known as types, but to avoid confusion with **type strains** (see below) the suffix 'var' is preferred. Thus, special biochemical or physiological properties distinguish biovars; serovars have distinctive antigenic characteristics; pathovars show pathogenicity with particular hosts; and phagovars (phage types) are lysed by certain bacteriophages. Although such infrasubspecific ranks have no standing in nomenclature they may be of great practical value; the recognition of over 2000 serovars of *Salmonella* for epidemiological tracing bears witness to this.

Genera and higher ranks

Each species must be assigned to a **genus** (plural, **genera**), and this is another rank without a satisfactory definition; it may be regarded as a well-defined group of one or more species that is clearly separated from other genera, and RNA relatedness studies are helping to clarify the concept in some cases. Additionally, it is not always clear which genus is the most appropriate home for a species but unless placed in one, however uncomfortable the fit, the species has no validity. Alternatively, a new genus might be created to accommodate it, depending on the lumping or splitting proclivities of the taxonomist.

The use of the formally recognized ranks higher than genus, namely families, orders, classes and divisions, is of limited value in phenetic classifications and only justifiable when there is evidence of genetic relatedness, as there is with phylogenetic schemes. The example of their use given in Table 1.1 is already obsolete, owing to the construction of phylogenies derived from sequence comparisons (see Chapter 6).

Table 1.1 The ranking of taxonomic categories based on the classification proposed by Murray*

Category	Example
Kingdom	*Procaryotae*
Division	*Gracilicutes* (the Gram-negative bacteria)
Class	*Scotobacteria* (the non-photosynthetic bacteria)
Order	*Spirochaetales*
Family	*Spirochaetaceae*
Genus	*Spirochaeta*
Species	*Spirochaeta plicatilis*

* *Bergey's Manual of Systematic Bacteriology* (1984).

Form of Names. The *International Code of Nomenclature of Bacteria* (Sneath, 1992) (also called *Bacteriological Code 1990 Revision*) regulates the form of names so that they are universally recognizable. The scientific name *B. anthracis*, and so the organism whose exclusive label it is, is recognized throughout the world; the same is not true of the organism's vernacular names which include bacilo do carbúnculo, Bactéridie de charbon, and Milzbrandbacillus. That said, the use of scientific names in communicating with persons familiar only with obsolete taxonomies and vernacular names can cause problems, as it has done for clinicians unaware that the plague bacillus is now called *Yersinia pestis* and not *Pasteurella pestis* or that *Salmonella typhimurium* is not the typhoid bacillus.

The scientific names of all taxa are distinguished by being Latin or latinized words, which are treated as Latin regardless of origin and which commonly have Latin or Greek roots. Each name has a definite position in the taxonomic hierarchy and the ranks above genus have specific endings such as '*-aceae*' for family and '*-ales*' for order.

The species name comprises two parts, the **genus name** followed by the **specific epithet**, and is known as a **binominal** (or **binomial**) **name** — a form popularized by Linnaeus. The genus name has a capital initial and the specific epithet does not, but both are italicized or underlined or made to stand out in some other way. Although the genus name can be used alone (when referring to the genus as a whole) and can be abbreviated as part of a binominal name when confusion with other genera possessing similar initials is unlikely, the specific epithet cannot. The word species may be abbreviated to sp. (plural, spp.) when referring to the taxonomic rank, but it is not a scientific name. For example, *Mycobacterium* (the genus), *Mycobacterium* spp. (several species of the genus), and *Mycobacterium bovis* or *M. bovis* (the species, assuming the latter will not be confused with *Mycoplasma bovis* for instance) are all permitted but *Mycobact.* and *Mycobact.* spp. are not, and nor is *bovis* as there are species with this epithet in at least six other genera. Obeying such a list of dos and don'ts may seem rather pedantic but it is not so very different to the conventions we observe, for much the same reasons, with our forenames and surnames.

Stability of names

Any taxon for which a name is proposed should be described in enough detail for it to be recognized by other workers, but in the early years of bacteriology, when there was no code of nomenclature, organisms were often inadequately characterized so that the names given to them were meaningless. Synonyms, therefore, abounded because investigators unable to recognize their new isolates from published descriptions frequently proposed new names.

The main aims of the *Bacteriological Code* are that names be stable, clear and meaningful, and to achieve these several principles are followed. Until the previous (1976) revision of the *Code*, however, the **principle of priority**, which states that the first name given to a taxon is the correct one, had become very difficult to obey, having inherited from the *Botanical Code* the starting date of 1753, when Linnaeus's *Species Plantarum* was published. The 1976 revision set a new starting date of 1st January 1980, by which time the majority of bacterial groups had been thoroughly overhauled by experts, most names discarded, and the *Approved Lists of Bacterial Names* could be published. The principle of priority is, therefore, now realistic and the other methods by which the aims of the *Bacteriological Code* are pursued can be considered.

Type cultures

The meanings of names are clarified by the establishment of **nomenclatural types**. At the species level these are reference specimens for names which are maintained as viable cultures in internationally recognized collections such as the National Collection of Type Cultures (NCTC) in London and the American Type Culture Collection (ATCC) in Washington, and when a new species is published the author must designate a **type strain** and deposit a culture in such a service collection. If the original type strain (the **holotype**) is lost or none was ever designated a **neotype** may be proposed. Cultures can then be bought by other workers and used for reference purposes in their taxonomic studies, or as controls in diagnostic laboratories. For those organisms that cannot yet be cultivated a preserved specimen may be the nomenclatural type.

The term type strain is misleading because there is no guarantee that the strain which chances to be so designated is typical of the taxon as a whole. Although it may appear to be typical when the taxon is first described, perhaps on the basis of very few strains, its designation is principally for nomenclatural purposes. It should only be regarded (once cultivated and subjected to a few confirmatory tests) as an authentic reference specimen; the most typical strain of a collection is called the **centrotype**.

As well as each species having a type strain maintained for it, one of the member species of a genus is named as the **type species** and may not be removed from that genus, and one of the genera in a family is designated the **type genus**. The name of the family is usually formed from that of the type genus (see Table 1.2).

Valid publication

Assuming that the preceding classification work is satisfactory and a synonym is not being created the **valid publication** of a new taxon must

Table 1.2 Some examples of nomenclatural types

Category	Taxon	Type
Family	*Vibrionaceae*	*Vibrio*
Genus	*Vibrio*	*Vibrio cholerae*
Species	*Vibrio cholerae*	National Collection of Type Cultures (NCTC) strain number 8021

include the following three elements: a new name that is different from that of any other microorganism, and in proper form; a description of the taxon's properties, especially those that distinguish it from other taxa and so make the proposition appropriate; and the designation of a type. All new names now have to be validly published in the *International Journal of Systematic Bacteriology* (*IJSB*) which is the official publication of the International Committee on Systematic Bacteriology of the International Union of Microbiological Societies and updates the *Approved Lists* regularly. Alternatively, a new name may be given **effective publication** in some other recognized journal, but its priority dates from a validating announcement in the *IJSB*. Names awaiting valid publication are given within quotation marks, as are the names of species which await transfer to different, perhaps new, genera because taxonomic studies show them to have been incorrectly assigned.

Nomenclature is necessary, even though the individual may regard it as a tedious chore, because without it the results of much painstaking classification work cannot be shared and used.

2 Phenotypic characters

A classification founded on any single character, however important that may be, has always failed; for no part of the organisation is universally constant.

Charles Darwin, 1859, *The Origin of Species*

Relationships

Before attempting to classify a group of bacteria it is important to be clear what kind of relationship is to be demonstrated. As explained in the previous chapter, bacterial classifications are usually based upon **phenetic relationships** which may be defined as *arrangement by estimates of similarity (resemblance) based upon a set of phenotypic characteristics of the strains under study*. This definition indicates, in reverse order, that the major steps in classification are: (i) collecting the strains; (ii) characterizing the strains; (iii) estimating similarities; and (iv) arranging the strains to show the relationships implied by their similarities. We may make our definition more precise by adding that the characters are equally weighted and that relationship by ancestry is not implied.

In fact, the rRNA sequence data upon which phylogenetic schemes for bacteria are based are also largely phenotypic but they reflect the genome faithfully enough to be regarded as genotypic (or **genomic**). If it is assumed that bacterial evolution has occurred at a steady rate and in an orderly fashion, the **cladistic** relationships can be deduced from similarity values which act as measurements of evolutionary time, and so branching networks (**cladograms**) like family trees can be drawn to show the most probable phylogenetic reconstruction.

Four main problems must be addressed when planning a phenetic classification of bacteria. The first two are that classifications based upon different character sets might show poor agreement (**congruence**), and the phenetic characters used represent a very small part of each organism's genome. These are combatted by basing a general purpose classification on a large number of characters from a wide phenotypic and genotypic range (and the **polyphasic** approach); that said, schemes founded on quite different character sets often show quite high congruence. The other two problems are concerned with analysis. Many methods exist for the calculation of similarities and for the arrangement of strains according to these similarities, and the application of different combinations of methods to a set of data can lead to a wide variety of interpretations. Fortunately, bacteriologists tend to restrict themselves to relatively few methods (those that happen to be available in the computer programs) and discrepancies between the results of different approaches may give useful insights into the stabilities and other aspects of the relationships implied.

Strains

The taxonomic term for any object to be classified is **operational taxonomic unit** (**OTU**) and for most bacteriological work the OTUs will be strains: throughout the remainder of this book therefore, the term strain will be used when discussing practical matters and OTU when considering the more theoretical aspects. Taxonomists can only be as good as their culture collections, so careful choice and handling of strains are essential to the success of a classification exercise.

Sources

For a general purpose classification the strains should be chosen to represent the known diversity and environmental niches of the notional group being studied. Narrower studies may quite legitimately concentrate on a group from a particular environment, marine *Bacillus* species for example, or an environment as a whole, such as the aerobic heterotrophic bacteria of an acidified loch; however, the former is too restricted to be regarded as a general purpose scheme and the latter is likely to include many disparate taxa and so may be too broadly based to give insights into the detailed structures of individual groups.

It is most important to include recent isolates because classifications based entirely upon 'museum pieces', which have been maintained in major culture collections and other laboratories for many years and have become adapted to laboratory conditions, lay poor foundations for diagnostic schemes intended to identify isolates from the real world. Notwithstanding this, type and other strains from culture collections are most useful as representatives of previously described species. They act as reference points in the study for nomenclatural purposes and facilitate comparison of the resulting classification with other schemes. It is therefore prudent to include several reference strains for each species in the study to give adequate coverage of within-taxon scatter and to allow for loss and death of cultures, or their not being what they are claimed to be.

Finally, a small proportion of the strains, perhaps 10%, should be chosen at random and duplicate cultures prepared for carrying through the study quite separately to permit the estimation of test error; this topic is discussed in more detail later.

These points are illustrated by a classification study which formed the basis of an identification scheme for the genus *Bacillus*. Over 1000 strains from soils, fresh and marine waters, salt marshes, man-made environments, plants, animals, dairy and other food products and their ingredients, soft and alcoholic drinks, beverages, biotechnological and other industrial plant and products, and medical and veterinary specimens were collected; about one-third of these were recent or fresh isolates, 8% were duplicates, and all species were represented by type and other reference cultures.

Numbers

One of the strengths of numerical taxonomy is that many OTUs can be handled. If the computer program to be used cannot cope with the full data set, a random sample of OTUs may be run; reference OTUs can then be chosen for each group formed and included in successive runs with the remaining OTUs. The results of these separate analyses can then be assembled. The main constraints are, therefore, how many strains can be obtained for study and, usually more important, how many can be adequately characterized if the project is to be concluded before support and enthusiasm wane. The number of strains needed for a general purpose classification depends on the tightness of the group being investigated, but the minimum probably lies between 50 and 100 and most studies use well in excess of these numbers. Furthermore, although it implies some prejudgment of groupings, it is desirable that each potential species be represented by at least 10 strains to give some indication of within-taxon scatter.

Maintenance

No matter how reliable the source *ought* to be, the purity of each strain must be confirmed on receipt and its identity checked with a few rapidly determined characters such as colonial morphology, cell shape and staining reaction. Such checks should be repeated periodically throughout the study and any doubtful cultures discarded. This requires that two collections be held, one of working cultures and the other a reserve stock.

Methods of maintenance will be dictated by the nature of the subject organisms, but all should aim to hold the strains in suspended animation, thus protecting against phenotypic variation as well as death, and yet permit ease of access, subcultivation where necessary, and storage. Reserves of many organisms may conveniently be kept in cryoprotectant at –80°C or as freeze-dried (lyophilized) cultures, with working stocks being regularly subcultured onto agar slopes or into broths which are stored at room temperature or 4°C after incubation.

Characters

Once the strains have been collected it is necessary to decide what properties are to be used in classifying them. Any property that can vary between OTUs is a **taxonomic character** and in so varying it will have at least two **character states** — for example, in a test for endospores they are positive (present) and negative (absent). Such **unit characters**, that we treat as independent because we cannot, with our present knowledge, divide them logically, may represent different amounts of the genome. The element of weighting that this introduces has to be ignored for practical purposes and its effect is largely 'ironed out' if many characters are used; ideally, each character would be determined by a single gene or operon. Some properties, on the other hand, may be very closely related — the end products of metabolism of one substrate for instance — and are referred to as **character complexes**. Obviously, the OTUs must be compared by corresponding or homologous characters that are applicable to most if not all of them. However, determining homology, to ensure that like is being compared with like, is rarely a problem in bacterial taxonomy.

The range of taxonomic characters is vast, but choice must be guided by several considerations including number, range and stability of characters, and the convenience and discriminatory value of the tests. The important subjects of test standardization and reproducibility will be discussed in a separate section.

Number

Experience indicates that the optimum number of characters to use lies between 100 and 150. Although the tests should give highly reproducible results this is hard to achieve in practice; the omission of all but the most reliable tests, however, could bring the number well below 100 but this would result in increased sampling error, and individual tests and test errors would exert disproportionately large influences on the estimates of similarity. In practice therefore, it is better to include as many tests as possible and omit only the very unreliable ones.

Theoretically, there is no upper limit on the number of characters that can be used, but to perform 200 or more tests requires a great expenditure of time and effort for a small return in increased accuracy. Furthermore, some tests might be inapplicable to certain strains and some types of characters could be over-represented.

Range

As a classification usually aims to reflect the many aspects of the subject organisms' biologies, it should attempt to represent the whole **phenome** (genotype and phenotype). The battery of tests, therefore, must be constructed from the widest possible range of properties, the various types of which should be evenly represented to prevent bias. The types of properties are often placed in nine categories as shown in Table 2.1. The last two types, serology and genomic characters, are sometimes difficult to score for numerical analyses and the measurements of reactions, such as percentages of DNA–DNA reassociations, may themselves be used as estimates of similarity.

Stability

The properties selected should be stable and little affected by slight environmental changes; chemotaxonomic characters are particularly dependent on environmental conditions, making strict standardization of test procedures essential. Also, strains often lose characters on repeated subculture and this is why fresh or recent isolates must be included in taxonomic studies.

Table 2.1 Categories of characters used in classification

Category	Examples
Cultural	Colonial morphology, pigmentation
Morphological	Cell shape, staining reactions, motility
Physiological	Growth temperatures, anaerobic growth
Biochemical	Acid from carbohydrates, nitrate reduction
Nutritional	Organic acids as sole carbon and energy sources, vitamin requirements
Chemotaxonomic	Amino acids in interpeptide bridges of cell wall, types of lipids in membranes
Inhibitory tests	Sensitivities to antibiotics, dye tolerance
Serological	Agglutination by antisera to reference strains
Genomic	mol%GC in DNA, DNA–DNA reassociation

Convenience

Given the large number of characters needed, it is wise to choose those that are easily determined and inexpensive, in time and money, to perform. Assuming their differential values, simple, cheap tests are also the best for any identification scheme that may be developed from the classification.

Discriminatory value

Tests giving uniformly negative or positive results are described as **redundant** because they have no discriminatory value. Redundant tests make no useful contribution to a classification and should be omitted if they can be recognized before characterization begins; it would be pointless to include Gram-negativity in a study of the *Enterobacteriaceae*, for example. Unfortunately, the redundancy of some tests will only become apparent late in the study. The more evenly a test separates the taxa, the greater its discriminatory value and such tests are important in the classification and useful for subsequent identification purposes. To select for highly discriminant tests at this stage, however, would be to prejudge the classification.

Test standardization and reproducibility

It has been said that a classification can only be as good as the culture collection on which it is based, but the potential of any collection can only be realized by high-quality data. Unfortunately, bacteriological tests are subject to considerable experimental error from several sources including biological variation, poor test standardization, and difficulties in reading results. It may be difficult for an individual worker to keep error below 2% in tests on duplicated strains, and results from different laboratories may show over 10% inconsistency so that the problem is exacerbated if any identification scheme developed from the taxonomy becomes widely used. Despite widespread awareness of the importance of test reproducibility, relatively little work has been done on the subject. As explained earlier, omitting all unreliable tests is not the solution; instead, the problem must be solved by rigorous test standardization and test error should be estimated in each study.

Test standardization

Strict standardization of all aspects of test procedure is essential. Factors which might affect the consistency of a test include: batch-to-batch variation of medium and reagent ingredients; variation in formulations of media and reagents; changes in procedures for sterilizing media and for cleansing of recyclable glassware; freshness of media and reagents; types of vessels and closures and tightness of closures; differences in sizes and ages of inocula; time, temperature, and gaseous conditions of incubation (the last two may vary within an incubator); poor instructions on interpretation (e.g. does nitrate reduction mean some or all nitrate reduced?) and variation in the use of descriptive terms; problems in the subjective interpretation of colour changes (in a red to yellow reaction, how yellow must an orange be to be scored positive?); unrealistic demands for recognition of many categories of reaction (e.g. slight differences in shape, or delicate nuances of colour) so that reading of results becomes tedious and drift in interpretation occurs; inaccuracy of measurement equipment.

It is clear that all test procedures must be exactly defined and closely adhered to throughout the work. Some tests, however, remain inconsistent when applied to particular bacterial groups; despite careful standardization they are just not amenable to improvement. Weakly reacting strains can also be a problem and test sensitivities and times of reading should take account of this. Strains known to give certain results in the various tests should be included as controls, but the reference cultures for the study will often largely serve this purpose.

Test error

If tests are poorly reproducible despite rigorous standardization, the data might be worthless and so some measure of test error is essential. This is conveniently obtained by duplicating about 10% of the strains at random and carrying them through the study blind under new code numbers. Test error may then be estimated from the data for the replicates using the equations of Sneath & Johnson (1972): individual test variances between duplicated strains (s^2) are calculated from

$$s^2 = d/2t$$

where d is the number of strains giving discrepant results and t is the total number of strains. Thus, in a study of 150 strains of which 10% are duplicated cultures, if six pairs give inconsistent results in a simple +/– test

$$s^2 = 12/300 = 0.04$$

For an individual test, **the probability of error** (P) is

$$P = 0.5[1 - \sqrt{(1 - 4s^2)}]$$

which in this case is 0.04 or 4%; tests with P>10% are best rejected. The effect of test errors on classification is considered in Chapter 4.

Characterization methods

This section outlines the different types of character categorized in Table 2.1 and considers their reproducibilities; methods more applicable to identification are covered in Chapter 5.

Cultural

Characters such as colony shape, margin, elevation, surface appearance, opacity, texture, pigmentation (enhanced, perhaps, by special media), odour, and appearance of growth in broth are rather unreliable because they are strongly influenced by medium, age of culture and incubation conditions and interpretation is somewhat subjective. Colony size is best omitted since it is an expression of growth rate and is greatly affected by competition from neighbouring colonies. It is important, therefore, that cultural characters are very carefully standardized and their reproducibilities assessed. That said, these characters are still useful in the hands of experienced bacteriologists and are especially valuable in identification.

Morphological

These characters include cell shape, curvature, size and arrangement (singly, chains, palisades, filaments), pleomorphism, formation of coccoid bodies, cysts, and other inclusions, spores (shape, position, size, swelling of sporangia), presence and arrangement of flagella, capsules, metachromatic granules, poly-β-hydroxybutyrate inclusions, and staining reactions such as Gram and acid fastness. The features are determined by phase-contrast microscopy of wet preparations, bright-field microscopy with cytological staining, and electron microscopy with shadowing. Some characters are often poorly reproducible: cell and spore shape suffer from worker subjectivity and bias (towards scoring as rods, for example), and staining of flagella can be difficult and tedious. The determination of cell size frequently causes problems, and eyepiece graticules, measurement of photomicrographs, and inclusion of latex beads of known sizes have all been found to be unreliable. Image-shearing eyepieces and eyepiece micrometers calibrated from stage micrometers give more consistent results, but even so the lengths of rods are bound to vary as they grow. It is now possible to obtain accurate measurements, and mean lengths and diameters for the cells in a microscopic field, by using an image-analysing microscope.

Motility. Motility may be determined by phase-contrast microscopy of wet preparations (of young cultures), the hanging-drop method, Craigie tubes, and migration through sloppy agar media. The best method to use is determined by the nature of the group to be studied, particularly its gaseous requirements, and results may be affected by incubation temperature. Special media and techniques are necessary for the observation of gliding motility.

Physiological

The maximum and minimum temperatures permitting sustained growth are most reliably tested for in liquid media incubated in waterbaths. Reproducibility is influenced by inoculum size and incubation time should be kept short for the higher temperatures, but at lower temperatures it may have to be prolonged. Temperature tolerance is an easily determined character but it is not widely used.

Growth at different NaCl concentrations may be tested using liquid or solid media. Measurement of pH range is less straightforward because the bacteria may vary in their resistances to different acidic ions. Also, media have to be well buffered and should contain indicators as the initial growth of the organisms may push the pH towards neutral.

Atmospheric requirements are best studied using solid media as gaseous conditions can vary with depth in liquid cultures and results from semi-solid media are inconsistent. Plates can be incubated in aerobic incubators, incubators or jars with 10% CO_2, incubators or jars with raised CO_2 and oxygen decreased to 5%, and anaerobic jars, to determine aerobic growth, enhancement of growth by raised CO_2, microaerophily, and anaerobic growth respectively.

Biochemical

The list of tests for specific enzymes and pathways is extensive, but only a limited number of those available is likely to be applicable to any one bacterial group and some organisms, the corynebacteria, for example, are relatively inactive in these tests. There is often a variety of media, reagents and procedures for a single test, each one developed for a particular bacterial group. Tests for acid production from carbohydrates, for instance, are dependent upon the buffering capacities of the media and the pH sensitivities of the indicators; a peptone-based medium containing Andrade's indicator or bromthymol blue, which have quite low pH end-points, is suitable for enterobacteria because they produce large amounts of acid, whereas an ammonium-salt-based medium containing an indicator such as phenol red, which has a fairly high pH end-point, is more appropriate for many non-fermenting organisms and *Bacillus* species which produce small amounts of acid that might be neutralized by alkali production in peptone media.

Reactions are often detected by incorporating indicators in the media but not all reactions can be revealed in this way and different indicators may inhibit certain organisms, be decolorized by others, or be obscured by the medium. Other reactions may be indicated by clearing of opaque media, precipitation and opacity in clear media, or by the addition of reagents to reveal the absence of a substrate or presence of a product, or by Durham tubes for trapping any gas evolved.

Biochemical tests can be divided into five groups.

1 Decomposition of simple carbohydrates: acid from glucose in aerobic and anaerobic conditions; gas from glucose; acid from many carbohydrates such as arabinose, fructose, lactose, mannitol, salicin, trehalose and xylose; methyl red; Voges–Proskauer test; β-galactosidase; and aesculin hydrolysis.

2 Metabolism of nitrogenous compounds: reduction of nitrate and nitrite; gas from nitrate; detection of end products including indole, H_2S and HCN; detection of urease, arginine dihydrolase, and decarboxylation of amino acids such as glutamine, lysine and ornithine, by alkaline reactions in media; phenylalanine and tryptophan deamination, producing keto acids which form coloured compounds with ferric ions; hippurate hydrolysis.

3 Decomposition of large molecules: the enzymes used include phospholipases (egg yolk reaction), lipases (tributyrin and Tween agars), proteases (gelatin and casein media), amylase (starch agar), cellulase, chitinase, DNase and hyaluronidase.

4 Terminal respiratory enzymes: catalase and cytochrome oxidase are tested for by rapid, non-cultural methods.

5 Miscellaneous: these include coagulase, phosphatase and haemolysis.

These tests were originally devised for identification purposes and many are now available in commercial kits that have been developed for several groups of bacteria. Most kits are for medically important organisms, especially the enterobacteria, *Staphylococcus*, *Streptococcus*, anaerobes, and non-fermenting Gram-negative bacteria, but some are for wider use with groups that include *Bacillus* (Fig. 2.1) and *Lactobacillus*. More versatile are kits for determining carbon source utilization, or those comprising enzyme tests applicable to a broad range of organisms and tissues. Even those kits aimed at specific taxa are often suitable for wider application. Although expensive to use for large studies, and not always sufficiently versatile, these kits do offer the advantages of convenience, miniaturization, rapidity (with automation in some cases) and, above all, strict standardization. None the less, with certain bacterial groups it can be difficult to obtain consistent results by certain methods for some of these 'classical tests' no matter how well standardized they are and be they in kit form or not: examples include gelatin liquefaction, nitrate reduction, and Voges–Proskauer tests.

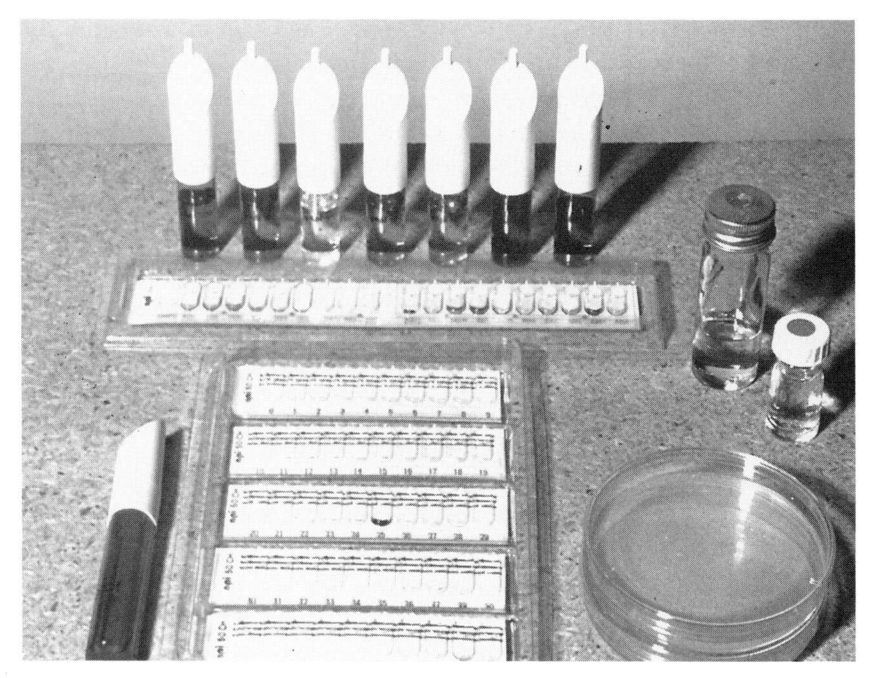

Fig. 2.1 The API test kit for *Bacillus* species. The strain is grown on the plates, harvested in distilled water, suspended in the test medium and distilled water to give a standard opacity, and inoculated into the test strips. After incubation, reagents are added, as appropriate, to reveal the reactions.

Nutritional

Bacteria show great diversity in the nutrients that they can utilize and a very wide range of substances, including carbohydrates, organic acids and amino acids, can be tested as sole sources of carbon, energy and nitrogen. Although some organisms such as certain *Pseudomonas* species are very versatile and utilize many different substrates, others may have complex requirements for vitamins and other growth factors and the investigation of these may be complicated. The defined basal media must be prepared from very pure ingredients and controls lacking the test substrate are most important — even distilled water or agar may contain growth-supporting contaminants. The more demanding organisms may need complex basal media which must be rigorously controlled. Test procedures should be highly standardized as result consistencies are strongly influenced by nutrient carry-over in inocula, size of inocula, concentrations and potential toxicities of substrates, incubation times and subjectivity in reading results. Also, certain subtrates and strains of some species may give particularly variable results.

A novel approach, commercially available as the Biolog system, is to use redox indicators such as tetrazoliums to detect colorimetrically the increased metabolic rate of a bacterial suspension which is oxidizing a carbon source or other substrate. Electrons donated to the electron transport chain by the NADH formed during oxidation, irreversibly reduce the colourless tetrazolium dye to a highly coloured formazan.

Chemotaxonomic

As shown in Table 2.2, many sensitive techniques can now be used to collect chemical information from bacteria. Such data are particularly helpful in the resolution of groups such as the actinomycetes, for which the traditional methods have failed to provide satisfactory classifications, and the

Table 2.2 Cellular components and methods in chemotaxonomy

Site or level	Components	Analytical methods*
Extracellular	Products of metabolism	TLC, GLC, HPLC and isotachophoresis of extracts and derivatives
	Polysaccharides	GLC of derivatives
Whole-organism	Whole-organism	FT-IR spectrometry
	Pyrolysate	Pyrolysis GLC or MS
	Fatty acids	GLC of methyl esters
	Proteins	PAGE, two-dimensional electrophoresis and isoelectric focusing
	Polyamines	HPLC of dansyl derivatives
	Sugars	GLC of methyl glycosides
Gram-negative outer membrane	Lipopolysaccharide	Purification by UC then GLC of methyl esters and glycosides
	Polar lipids	TLC of solvent extracts
Mycobacterial outer membrane	Mycolic acids	TLC of methyl esters
	Free lipids	TLC of solvent extracts
Cell wall	Peptidoglycan:	
	1 Diaminopimelate isomers	PC and TLC of whole organism or wall hydrolysates
	2 Other amino acids	PC and TLC of wall hydrolysates, GLC of derivatives, AAAA
	3 Glycan type	Glycolate test
	Polysaccharides	As for whole-organism sugars
	Teichoic acids	Chemical analysis and serology
Plasma membrane	Isoprenoid quinones Polar lipids	TLC, RPTLC, GLC, HPLC, RPHPLC and DPMS of solvent extracts
	Cytochromes	Spectrophotometry
Proteins	Enzymes	PAGE, growth independent enzyme assays (rapid enzyme tests) and functional characterization
	Amino acids	Sequencing

* Abbreviations: AAAA, automatic amino acid analysis; DPMS, direct probe mass spectrometry; FT-IR, Fourier-transform infrared; GLC, gas–liquid chromatography; HPLC, high-performance liquid chromatography; MS, mass spectrometry; PAGE, polyacrylamide gel electrophoresis; PC, paper chromatography; RPHPLC, reverse-phase HPLC; RPTLC, reverse-phase TLC; TLC, thin-layer chromatography; UC, ultracentrifugation.

archaea. The characters are useful at all levels in the taxonomic hierarchy and, given standardized conditions of cultivation and preparation, they are highly stable. The discriminatory powers of the various chemical criteria differ between groups; peptidoglycan type, for example, is quite uniform in Gram-negative bacteria, and so of limited taxonomic interest, but very variable and taxonomically valuable in Gram-positive genera. Furthermore, the taxonomic significances of chemical differences probably vary — substitutions by chemically similar components may be less significant than those by very dissimilar ones.

Most procedures involve separation by chromatographic or electrophoretic techniques, but until relatively recently results were often interpreted visually, owing to problems of reproducibility and data handling, rather than being subjected to numerical analysis. Analytical methods vary in precision, speed and convenience. Some of the more sophisticated techniques can handle only a few strains at a time whereas the simpler approaches, which may furnish cruder, less resolved, qualitative data, can handle many and so yield more comprehensive taxonomies. When only small numbers of representatives can be processed they should be chosen carefully, with reference to classifications based upon other characters.

Many of the traditional biochemical tests such as methyl red and Voges–Proskauer provide information about metabolic end products, but chromatographic analyses yield quantitative data on the different fatty acids, alcohols, aldehydes, ketones and amines that may be produced and these data can now be directly processed by computer. This approach has chiefly been used in the classification of anaerobic bacteria where the patterns of fatty acids from carbohydrate and protein metabolism are of particular value.

Standardization. A major constraint in chemotaxonomic work is the dependence of bacterial chemical composition on the environment. Some components, such as peptidoglycan, have very stable compositions in most organisms but may sometimes be affected by abnormal growth conditions; for example, the amino acid balance of the medium can influence the composition and structure of *Staphylococcus* peptidoglycan. Other components and products are strongly influenced by changes in environmental conditions; fatty acids are particularly sensitive (those in the plasma membrane affect membrane fluidity and their alkyl chain lengths are increased at higher temperatures). Also, differences in end-product profiles for anaerobic organisms can be at least as great for a single strain grown on different media as for different strains grown on the same medium, and even a single strain grown on separate batches of the same medium might yield inconsistent results. It is therefore of paramount importance that media, incubation conditions, and time of incubation are rigorously standardized in all chemotaxonomic studies, especially quantitative ones, and this may inhibit the wider application of these methods to groups of organisms with different growth requirements.

It is also most important to standardize the preparative procedures and analytical methods wherever possible. In polyacrylamide gel electrophoresis (PAGE) of proteins, for example, reproducibility is not only affected by cultural conditions, but also by cell disruption methods, ages of samples, gel structure and running conditions, and several gels must be run to verify results. Gradient PAGE, two-dimensional PAGE and isoelectric focusing methods give greater resolution, allowing the recognition of several hundred bands compared with about 30 from PAGE, but these also bring greater problems with reproducibility.

Pyrolytic methods (in which cells are disrupted by high temperature in an inert atmosphere) require careful control of pyrolysis parameters — the time taken to reach the set temperature and the speed with which the products are removed for analysis to avoid any secondary reactions. The reproducibility of pyrolysis gas–liquid chromatography (PGLC) suffers from column ageing and inconsistency between columns, and even pyrolysis mass spectrometry (PMS) equipment can be subject to drift. Furthermore, single components, such as polysaccharide capsular material produced in profusion, may dominate spectra so that variables become unequally weighted.

Although problems such as the above might seem to limit the wider use of some chemotaxonomic methods, if standards are included corrections often may be made during computer analysis (see Fig. 2.4 later).

Lipids. The non-hydroxylated fatty acids with chain lengths of up to 20 carbon atoms, which are found in plasma membranes and lipopolysaccharide, are taxonomically very useful and can be analysed by GLC of their methyl esters (FAMES) (Fig. 2.2). Most eubacteria (now called the *Bacteria,* one of the two domains of prokaryotes, the other being the *Archaea,* see Chapter 6) can be divided into two broad groups according to whether they possess mainly straight or branched chain fatty acids; other features include chain length, unsaturation and cyclopropane substitution, and hydroxylation of lipopolysaccharide fatty acids, so giving some taxa characteristic fatty acid profiles. The mycolic acids, which are 3-hydroxy-2-branched fatty acids with chain lengths of 24–90 carbon atoms, have only been found in *Corynebacterium*, *Mycobacterium*, *Nocardia* and closely related taxa and are believed to be associated with acid fastness. The discontinuous distribution of these complex lipids, which show wide structural variation, is valuable at the genus level and, in the case of *Mycobacterium*, at the species level as well.

Polar lipids are essential components of plasma membranes, playing important roles in regulation and permeability. The commonest types are phospholipids, several of which — phosphatidylglycerol (PG), diphosphatidylglycerol (DPG), and phosphatidylethanolamine (PE) — are widely distributed, and are of little taxonomic value for most Gram-negative groups. Others are highly characteristic of certain taxa and overall patterns are often useful. In actinomycete taxonomy, for example, patterns ('fingerprints') from two-dimensional thin-layer chromatography, the distribution

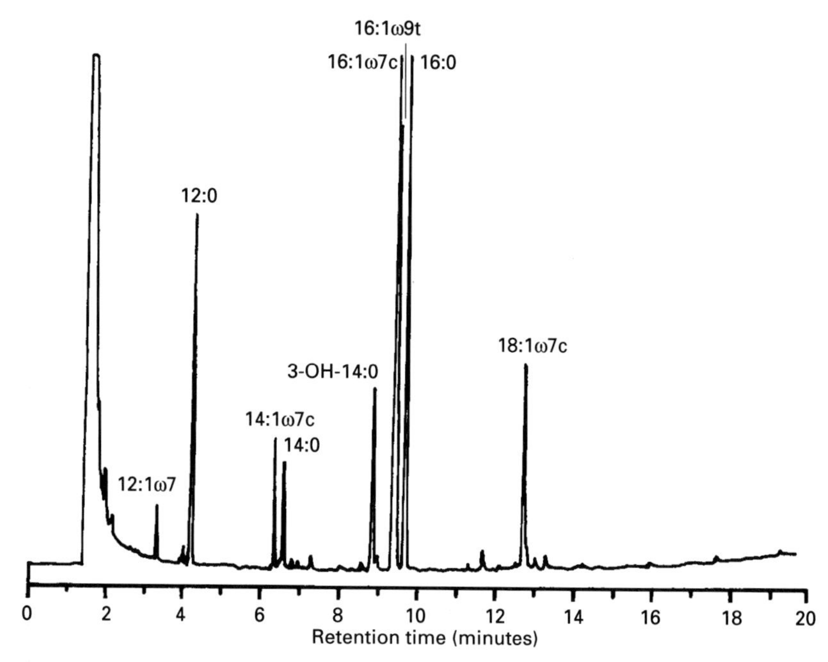

Fig. 2.2 Gas chromatogram of methylated fatty acids of *Campylobacter cryaerophila* (now *Arcobacter cryaerophilus*). This organism contains two monounsaturated fatty acids absent from other *Campylobacter* species, namely C14:1ω7c and C16:1ω9t (shorthand designations indicate number of carbon atoms: number of double bonds, and position of double bond from the hydrocarbon end of the *cis* or *trans* isomer).

of the common markers PG, DPG and PE, and the recognition of characteristic types such as phosphatidylinositol mannosides are very useful. Another common group of polar lipids, the glycolipids, are widely found in Gram-positive bacteria and are of some taxonomic value. Archaeal lipid structures are based on ether-linked branched chain types unique to this domain.

Isoprenoid quinones are found in the plasma membranes of aerobic, facultatively anaerobic, and some strictly anaerobic bacteria, where they are important in electron transport and oxidative phosphorylation. The type of isoprenoid quinone may be determined by thin-layer chromatography (TLC) and the individual components separated by reverse-phase TLC (in which the stationary phase is polar and the mobile phase non-polar) or, for quantitative assay, reverse-phase high-performance liquid chromatography. The main types, menaquinones and ubiquinones (Fig. 2.3), exhibit ranges of isoprenologues, as their side chains have lengths varying between 1 and 15 isoprene units; ring demethylation and degree of side-chain hydrogenation also vary in menaquinones so that this type is of greater discriminatory value. In facultatively anaerobic Gram-negative genera menaquinones, demethylmenaquinones, and ubiquinones are found alone or in various combinations, but most strictly aerobic genera produce only ubiquinones. Cytophagas and myxobacteria, however, produce only menaquinones, as do many species of the strictly anaerobic genus *Bacteroides* and

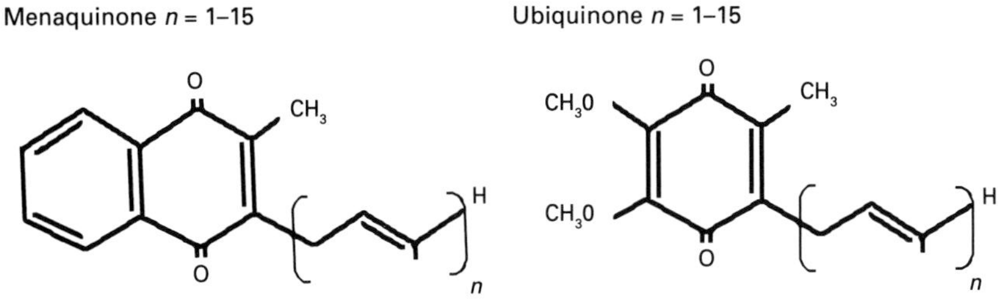

Menaquinone n = 1–15   Ubiquinone n = 1–15

Fig. 2.3 Structures of menaquinones and ubiquinones.

those mollicutes (or mycoplasmas) and aerobic archaea examined to date. Menaquinones are also the main types found in aerobic and facultatively anaerobic Gram-positive genera and they are of considerable value in the classification of some cocci and certain actinomycetes. The isoprenoid quinones of cyanobacteria are, interestingly, of types found in plants.

Proteins. The genetic codes for cellular proteins represent a large part of the bacterial genome and under strictly standardized conditions closely related bacteria will have similar protein contents. These may be analysed by electrophoresis of cell-free extracts or, where greater discrimination is required, by comparisons of specific proteins using enzyme activity studies, serology, and amino acid sequencing. Polyacrylamide gel electrophoresis (PAGE) of whole-organism proteins is a powerful, cheap and relatively simple technique. Methods such as two-dimensional PAGE can detect single protein differences, but such high resolution is not usually needed in systematic studies and fewer strains can be handled at one time. Proteins are usually released and solubilized by detergents such as sodium dodecyl sulphate but other chemical or mechanical methods of cell lysis may be required. Gels are stained to reveal the protein bands and, with automatic scanning densitometry and direct computer processing of data, this method is considerably faster than DNA–DNA reassociation and gives equivalent information. Another advantage is that a data set is obtained for each strain, whereas the DNA method is based on comparisons with reference strains. The results of PAGE are therefore amenable to numerical analysis (Fig. 2.4) and can be used to establish databases for identification. Despite these advantages, and good correlation with the results of traditional methods and DNA–DNA reassociation studies, protein electrophoresis has not been as widely adopted as it deserves to be.

Amino acid sequencing of specific proteins, quantitative serology and other techniques such as spectrophotometry of cytochromes yield information of value for determining phylogenies, the assumption being that most extant proteins have evolved from a few archetypal ones. These approaches give some useful insights and although eclipsed for a time by rRNA analyses they have become widely employed in phylogenetic studies.

Polyamines. Polyamines are found in all cells and are polycationic compounds that are primarily synthesized from amino acids and contain two

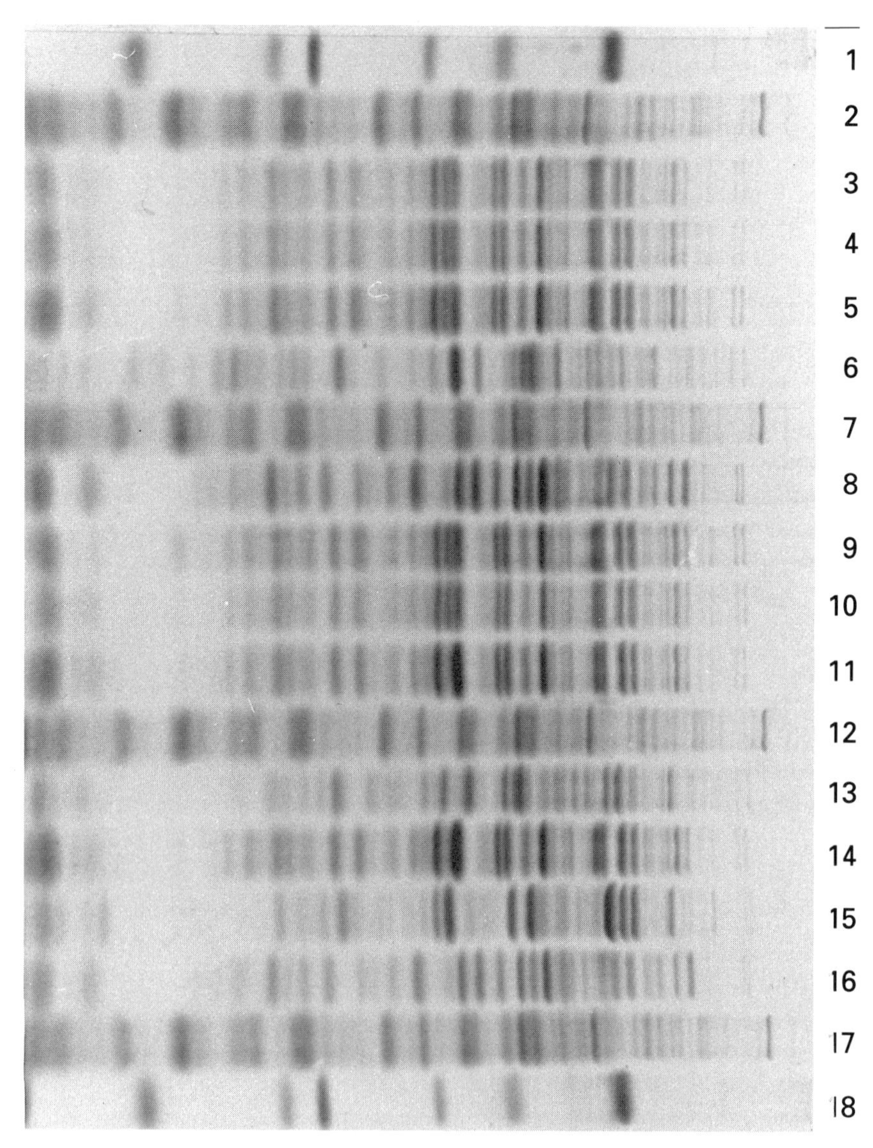

Fig. 2.4 A PAGE gel showing the heterogeneity of protein patterns obtained when strains of the genera *Lactobacillus*, *Lactococcus* and *Enterococcus* are compared: lanes 3–5, *Lactobacillus paracasei*; lane 6, *Lactococcus lactis* subsp. *lactis*; lane 8, *Enterococcus faecium*; lanes 9–11, *Lactobacillus paracasei*; lane 13, *Lactobacillus plantarum*; lane 14, *Lactobacillus paracasei*; lane 15, *Lactobacillus sake*; lane 16, *Enterococcus faecium*. Lanes 1 and 18 are molecular weight markers, and lanes 2, 7, 12 and 17 are *Psychrobacter immobilis*, used as a reference to allow the patterns in the corresponding areas of the gel to be normalized during computer analysis.

or more amino groups. Their function is not clear but they appear to have roles in DNA replication, tRNA and cell membrane stability, rRNA structure and protein biosynthesis. While putrescine and spermidine are widespread, other polyamines including cadaverine and spermine are less common, and some such as norspermine and 2-hydroxyputrescine are quite unusual. Qualitatively and quantitatively therefore, they are useful

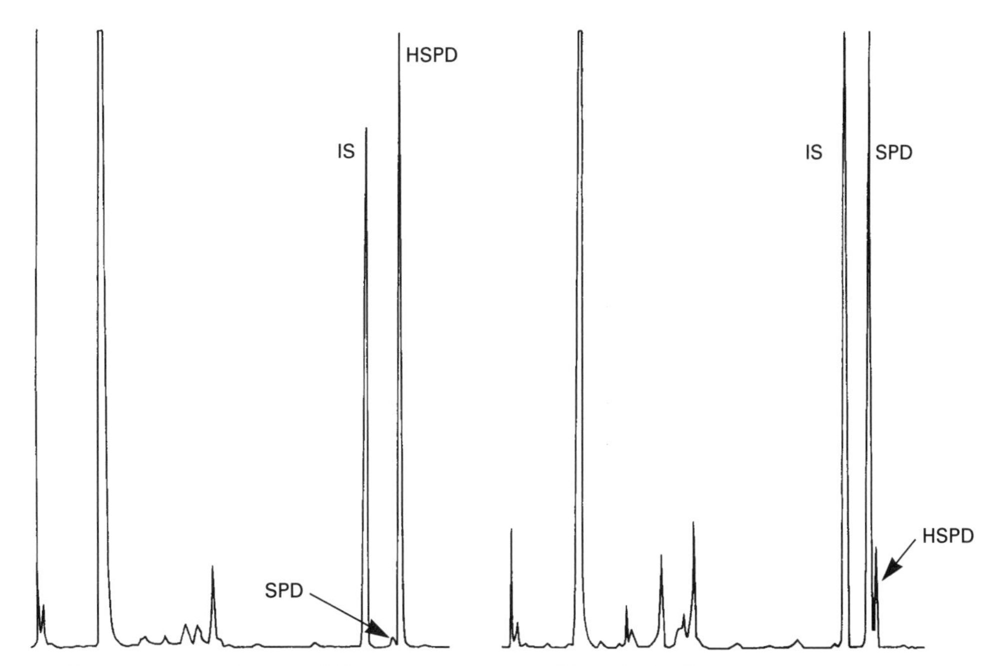

Fig. 2.5 Chromatograms showing different polyamine profiles which allow the differentiation of the type strains of two species incorrectly assigned to *Pseudomonas:* a " *P. diminuta*" and b "*P. vesicularis*". HSPD, homospermidine; IS, internal standard; SPD, spermidine.

chemotaxonomic markers and they have been investigated in a number of groups including *Agrobacterium, Bacillus, Flavobacterium, Halococcus, Pseudomonas, Rhizobium, Vibrio* and the methanogens and other archaea.

Cells are grown on a standardized medium, as the nitrogen components in particular must be controlled, and harvested at a standard time. The polyamines are extracted using hot perchloric acid and derivatives obtained using dansyl chloride. The dansyl derivatives are then separated by high-performance liquid chromatography (HPLC) (Fig. 2.5).

Cell Walls. Most of the eubacteria possess the characteristic cell wall polymer **peptidoglycan** (or **murein**) (Fig. 2.6), the mollicutes, chlamydiae and some budding bacteria being notable exceptions. It is composed of alternating *N*-acetylglucosamine and *N*-acetylmuramic acid units forming a glycan chain which is cross-linked by peptide binding between side chains that are commonly of the sequence L-alanine, D-glutamate, an L-diamino acid and D-alanine. In Gram-negative organisms cross-linking usually occurs directly between amino acids at positions 1 and 3 in adjacent tetrapeptides, but in Gram-positive bacteria it is usually by interpeptide bridges of 1–6 amino acids between the carboxyl group of the D-alanine at position 4 of one side chain and either the free amino group of the diamino acid at position 3 (group A peptidoglycans, Fig. 2.6a) or, sometimes, the

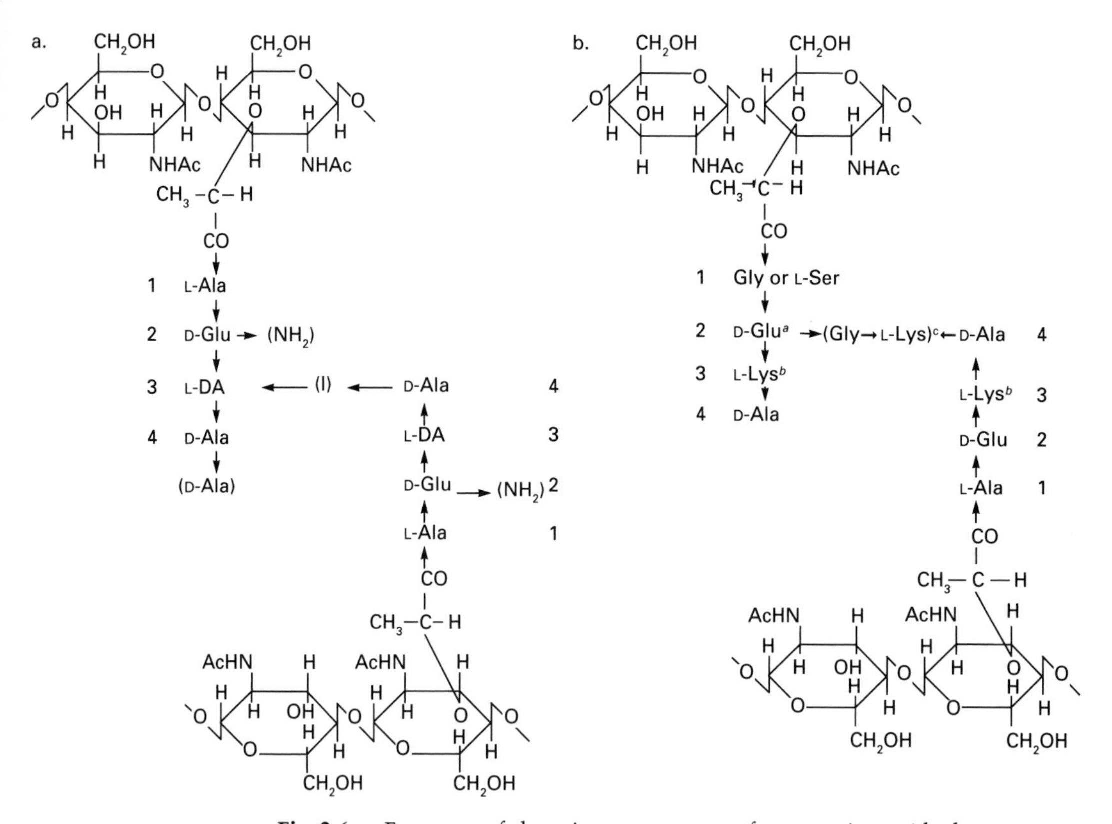

Fig. 2.6 a. Fragment of the primary structure of a group A peptidoglycan. Abbreviations: Ac, acetyl or, rarely, glycolyl; L-DA, diamino acid; I, interpeptide bridge; parentheses indicate items which may be absent.
b. Fragment of the primary structure of a group B peptidoglycan.
Notes: [a] sometimes hydroxylated; [b] or L-Glu, L-Ala, L-Orn, L-Hsr, L-Dab; [c] in subgroup 1, in subgroup 2 bridge may be D-Lys, or D-Orn, or Gly→D-Orn, or D-Dab.

carboxyl group of the D-glutamate at position 2 via a diamino acid such as lysine (group B peptidoglycans, Fig. 2.6b), in the other chain.

The presence of optical isomers of the widely distributed diamino acid 2,6-diaminopimelic acid (DAP) at position 3 in the tetrapeptide is easily detected at the whole-organism level; for determination of other amino acids at this position and elsewhere in the tetrapeptide, and those in the interpeptide bridges, quantitative amino acid analysis and sequencing of purified peptidoglycan are required. Gram-negative bacteria produce only small amounts of a uniform type of peptidoglycan, but in Gram-positive organisms variations in the tetrapeptide side chains and the interpeptide bridges provide valuable chemotaxonomic markers which have been especially useful in the classification of actinomycetes.

Bacterial cell walls contain various sugars in addition to the glucosamine and muramic acid of the peptidoglycan and although not all taxa possess uniform sugar compositions, certain groups, especially actinomycetes and some other Gram-positive bacteria, have taxonomically useful cell wall sugar patterns which can be simply determined at the

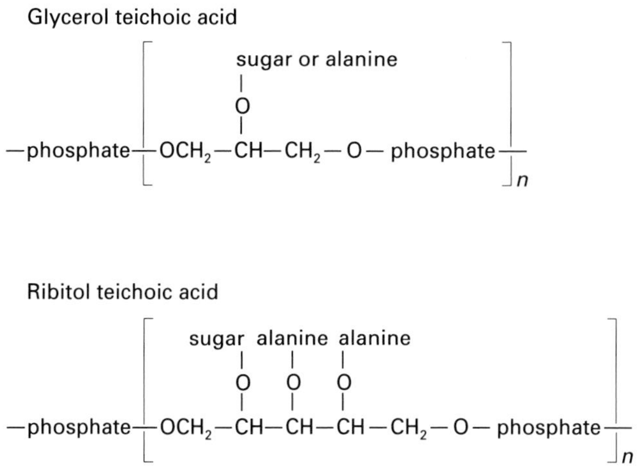

Glycerol teichoic acid

sugar or alanine
|
O
|
—phosphate—[OCH_2—CH—CH_2—O— phosphate]$_n$—

Ribitol teichoic acid

sugar alanine alanine
| | |
O O O
| | |
—phosphate—[OCH_2—CH—CH—CH—CH_2—O— phosphate]$_n$—

Fig. 2.7 Teichoic acid structures.

whole-organism level. The glycan moiety of the peptidoglycan is fairly uniform, but in some mycolic acid-containing actinomycetes the *N*-acetylmuramic acid is substituted by *N*-glycolylmuramic acid, which can be detected by a simple colorimetric test, and this determination of cell wall acyl type is of potential taxonomic value.

Gram-positive bacteria also contain large amounts of teichoic acids. These are water-soluble phosphodiester-linked polymers of glycerol phosphate, ribitol phosphate, or other polyol phosphates, which may be substituted by sugars, amino-sugars, or D-alanine (Fig. 2.7). They occur as cell wall teichoic acids, covalently bound to peptidoglycan, and lipoteichoic acids associated with the plasma membrane; the diversity of the former type has been most helpful in *Staphylococcus* taxonomy.

The general structure of Gram-negative bacterial lipopolysaccharide is fairly uniform but the composition and internal arrangement of sugars and fatty acids is very variable. Despite the availability of sensitive chromatographic techniques, chemotaxonomic investigations of lipopolysaccharide, and of capsular polysaccharides, have been rather few, with serological approaches being more popular.

Serology

The basis of comparative serological studies is that specific antigenic components of one organism react more strongly with an antiserum raised against the antigens of a similar organism than with one raised against the antigens of a dissimilar organism. A very wide range of immunological techniques based upon agglutination, precipitation, neutralization, immunofluorescence, and radio- and enzyme-immunoassay may be used but strict standardization and quantification are essential. There are many sources of error in serological tests including variations in dilution media, pipetting methods, substrates (e.g. cell cultures) and incubation conditions; intrinsic factors such as purity of the antigen, the animal strain used for antiserum production, the proportions of antibody types in an antiserum, antibody specificity, antigen/antibody ratios and the nature of the reaction also affect reproducibility.

For some bacterial groups, the enterobacteria for example, the antigenic complexities of the cell surfaces result in numerous cross-reactions and permit the recognition of so many serovars that objective analysis is practically impossible; such fine divisions are very valuable in epidemiological studies but of limited use in classification. With groups having less complex antigenic complements serological data may be useful for confirming the findings of classifications based upon other characters, provided that a reference antiserum is available for each main taxon.

Data analysis may be difficult because of the asymmetry of cross-reactions: the reaction of strain A with antiserum raised against strain B may be stronger than the reaction of B with antiserum raised against A. Although several numerical methods for comparative serology are available they have been little used.

The use of quantitative serology in chemotaxonomy has already been mentioned. Such an approach is limited to the comparison of proteins with relatively high homologies and looks only at their surfaces, which are their least stable parts. None the less, this method is relatively rapid and convenient compared with full structural analysis and has been of some use in bacterial classification; the study of proteins with differing evolutionary rates can yield phylogenetically valuable information.

Pyrolysis. Pyrolysis methods indirectly examine the total cellular compositions of bacteria, yet despite advances in chemical interpretation there is still a lack of understanding about what pyrolysis gas chromatograms and mass spectra represent. Although most bacteria produce roughly similar pyrolysate patterns, with multivariate computer analysis these can be distinguished on the basis of their peak heights, which show reproducible differences between taxa (Fig. 2.8) even at the strain level. Interpretation can be difficult with this and other such 'fingerprinting' methods, as it is not always clear what taxonomic levels are being revealed, and sophisticated statistical methods are needed to separate taxonomic information from 'noise'.

Of the few published classification studies based upon pyrolysis data, most have been technical rather than taxonomic investigations, but they show good correlation with schemes derived by other methods. As already indicated, PMS instrument instability can be a problem in long-term characterization studies, be they for classification or identification, and a corrective algorithm calibrated by a control strain related to the group under study may need to be applied. Another problem is that only a few species can be handled at one time, otherwise the numbers of discriminatory peaks become diluted. Cost of equipment has been an obstacle to widespread use of PMS, but systems continue to become more affordable, and as the method is very rapid and may be applied to any cultivable organism its chemotaxonomic use deserves further development.

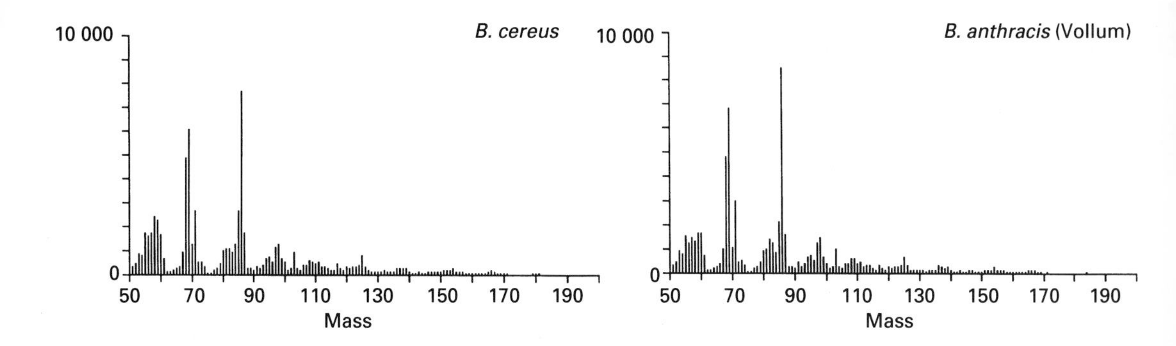

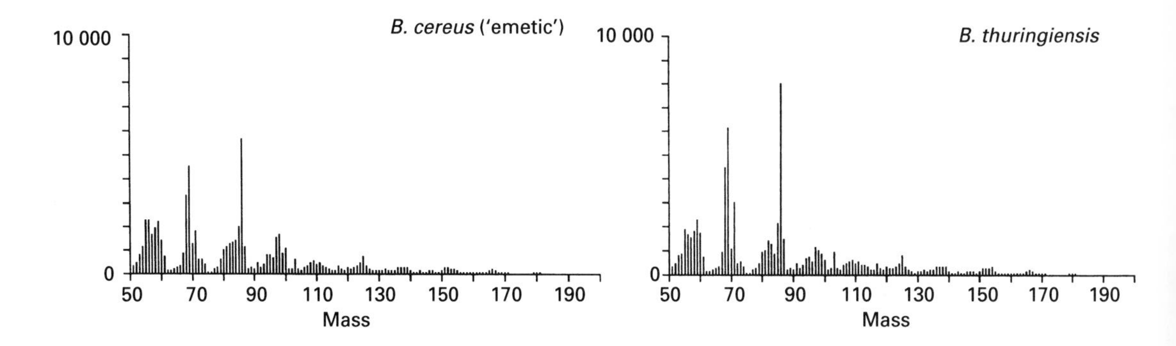

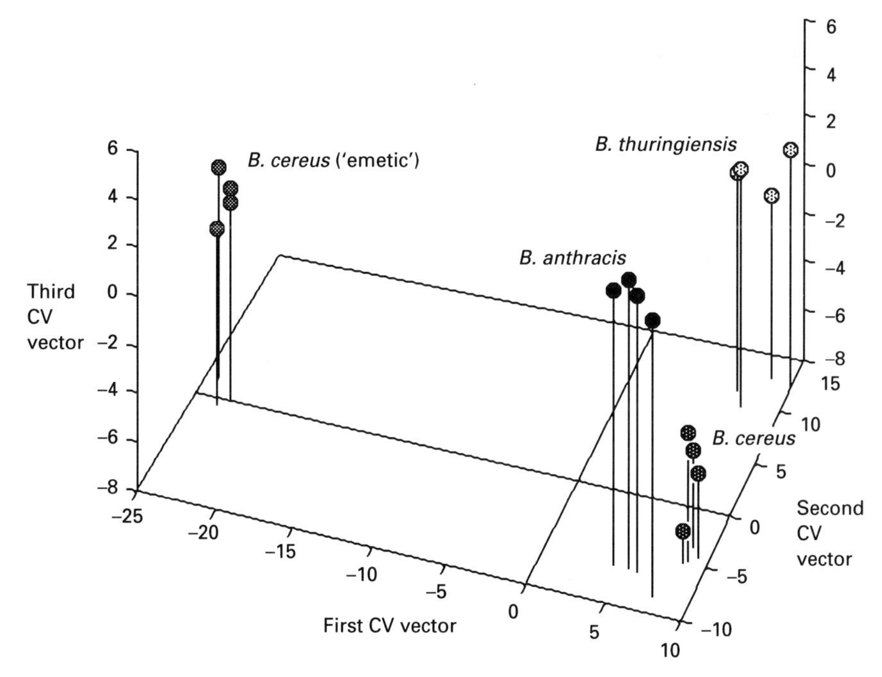

Fig. 2.8 Pyrolysis mass spectra of four closely-related *Bacillus* strains, showing the similarity that usually exists between spectra of close relatives, and a three-dimensional plot of a canonical variates analysis of the kind of data shown in the four spectra. 'Emetic' strains of *B. cereus* represent the biovar most commonly associated with outbreaks of the emetic type of food poisoning.

Fourier-Transform Infrared Spectrometry and Ultraviolet Resonance Raman Spectrometry. Fourier-transform infrared spectroscopy of a transparent film of cells yields a complex spectrum of overlapping absorbance bands representing the organism's total chemical composition. The Fourier-transform is a function that relates a complex waveform to its component spectrum of frequencies. Some parts of the spectrum are dominated by certain cell components and the contours of these more specific areas can be compared by correlation analysis. This rapid and convenient approach is claimed to be useful for classification and identification from serogroup up to genus level, but is not widely used.

Ultraviolet resonance Raman spectroscopy is another potentially powerful method for differentiation of bacteria. Monochromatic light is scattered by the specimen and interactions result in wavelength shifts (the Raman effect); it has been shown that ultraviolet excitation of bacteria at 222 nm and 231 nm produces spectra giving information on cell wall and protein components respectively.

Inhibitory tests

Although bacteria may easily acquire resistance to antimicrobial agents, in many groups, especially those of little medical or veterinary importance, sensitivity patterns are fairly constant and of value for classification and identification. Sensitivities to antibiotics, chemotherapeutics and other antimicrobials such as the vibriostatic agent 0/129, have frequently been tested in numerical taxonomic studies. Disk diffusion is the simplest and most popular method; results may be scored quantitatively (radius of inhibition zone) and are satisfactorily reproducible given fresh commercially prepared disks and carefully formulated and standardized media.

Tolerance of dyes such as brilliant green, crystal violet and fuchsin, of heavy metals such as cadmium, lead and mercury, and of various toxic or inhibitory salts including potassium tellurite, sodium azide and thallous acetate, have been used in several studies; more commonly growth on selective media which may contain one or more such agents is determined — examples are eosin methylene blue agar, bismuth sulphite agar and deoxycholate citrate agar.

3 Genotypic characters

Watson and Crick had discovered that formation of complementary hydrogen bonds seems to be all there is to the process by means of which like begets like.

Gunther S. Stent, 1981, Introduction to *The Double Helix*

Introduction

Unlike other cell constituents used for chemotaxonomy, only the amounts and not the compositions of RNA and chromosomal DNA are affected by growth conditions. Furthermore, the nucleic acids are universally distributed and they alone can be used as standards for wide-ranging comparisons.

The richest sources of information in the nucleic acids are the nucleotide sequences of complete bacterial genomes, and these are, of course, the theoretical goals of all characterization methods. As routine sequencing of complete genomes is not feasible at present, several alternative approaches are taken. They include estimating the mean overall base composition of DNA, comparing nucleotide sequences by DNA–DNA pairing studies, generating unique sets of DNA fragments by digestion with restriction endonucleases, sequence comparisons of selected genes, DNA–rRNA hybridization, and sequencing of rRNA.

Preparation

The main drawbacks of nucleic acid studies are the tedious preparation and purification procedures required, but for small quantities autoextractor machines have now been developed, and the polymerase chain reaction (PCR) may then be used to amplify those parts of the genome of interest.

For large amounts, organisms are grown in conditions giving high yields of cells and up to 2–4 g of washed cells are then lysed with sodium dodecyl sulphate or other detergent. Gram-positive cells require initial digestion with lysozyme and members of some genera such as *Mycobacterium* and *Streptococcus* may need pretreatment to make them susceptible to it. The centrifuged lysate is then deproteinized enzymatically or by treating repeatedly with chloroform or phenol, and the nucleic acids are extracted by precipitation in ethanol. The unwanted nucleic acid is removed by treatment with RNase or DNase as appropriate and the nuclease and any residual proteins are removed by treatment with a non-specific protease such as pronase. DNA, which must be free of protein, carbohydrate and RNA impurities, may then be selectively precipitated by isopropanol. In all work with RNA strict precautions must be taken to render all materials free of contamination by RNases.

Polymerase chain reaction

The polymerase chain reaction allows the *in vitro* replication of defined DNA sequences, and so it is of great value in bacterial classification and identification. Examples of its application are the genotypic analysis of non-cultivable organisms (which represent the majority of prokaryotes), and enhancement of gene probe detection of specific organisms in complex specimens such as environmental and clinical samples, and symbiotic associations.

The technique involves numerous repetitions of a three-stage process (Fig. 3.1) which may be carried out in an automated thermal cycler:

1 Relatively pure double-stranded (ds) DNA is **melted** (denatured by heating) at about 95°C to give single-stranded (ss) DNA;

2 Extension primers, which are pairs of ss sequences of typically 18–28 nucleotides complementary to the ends of the target sequence, are attached

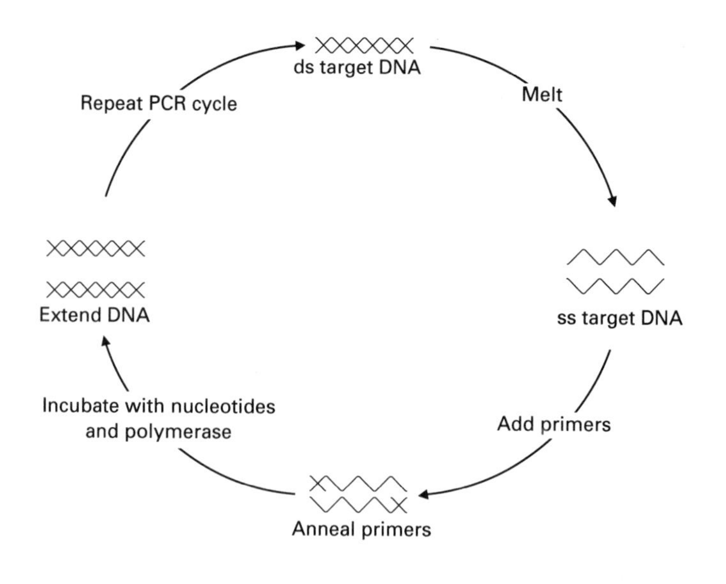

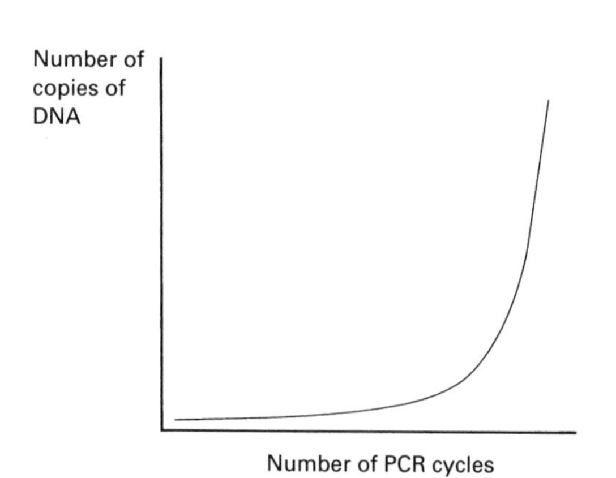

Fig. 3.1 The stages of the polymerase chain reaction (PCR) and the resulting amplification of copies of the target region of DNA.

to sites flanking the region to be replicated. This primer annealing takes place at a temperature 5°C below the **melting temperature** (T_m) of the primers, in the range 55–72°C;

3 The primers are extended along the target sequence by adding nucleotides using DNA polymerase. The *Taq* DNA polymerase most commonly used is derived from the thermophile *Thermus aquaticus*. It is thermostable, allowing it to survive the melting stage, and has a temperature optimum of 72°C.

The target sequence is thus duplicated (though not without some errors), and the cycle may be repeated perhaps 50 times or more in order to amplify the number of copies exponentially.

DNA analysis

Base composition

Among prokaryote DNAs the ratio of the nucleotide bases adenine (A) plus thiamine (T) to guanine (G) plus cytosine (C) ($[G + C] / [A + T + G + C] \times 100$; expressed as mol% G+C) varies within the range 23–78%. The greater the disparity in mol% G+C between two organisms the less closely related they are; theoretically, DNA molecules with differences of greater than 20–30% G+C can have virtually no sequences in common. Guidelines suggest that organisms showing more than 10% difference should not be assigned to the same genus and that 5% is the maximum range permissible for a species. However, experimental findings do not always support this theory. Some groups such as *Treponema*, and some of mollicutes, which are closely related according to other genomic methods, show G+C ranges close to 20 mol%. The factors which cause changes in G+C contents are not understood and the rates at which such changes occur are not known. Furthermore, estimates must be treated with caution: no DNA preparation shows absolute molecular homogeneity so that G+C content is always a mean value, and determinations made in different laboratories may disagree (for the type *strain* of *Bacillus subtilis* a range of 5.7% from four determinations has been reported).

It must be appreciated that although differences in mol% G+C are taxonomically useful for separating groups, similarities in base compositions do not necessarily indicate close relationships because the determinations do not take the linear sequences of bases in the DNA molecules into account; the criterion can only be used negatively. Thus, although mol% G+C cannot be used like other taxonomic characters, it is a very valuable indicator of the homogeneity of a taxon and so is usually quoted in any description of a group.

Determination. Direct estimation by hydrolysis and quantitative chromatography is the basis for the empirical formulae used in the other, physicochemical, methods but it is not itself used routinely. The two methods commonly used are relatively insensitive to RNA and protein contamination and are based upon the thermal stability and density of the molecule — the greater the proportion of G+C the higher these are.

The thermal denaturation of dsDNA into ssDNA, by breakage of the hydrogen bonds that bind the two strands, is accompanied by an absorbance increase of about 40% at 260 nm. There are three hydrogen bonds between G and C but only two between A and T, and the T_m at which this hyperchromic shift occurs is linearly related to the mol% G+C (Fig. 3.2). This method is popular because it is rapid and requires only an ultraviolet spectrophotometer with a heated cell. The other method is less popular because an analytical ultracentrifuge is required for long periods. High-speed centrifugation of a caesium chloride solution yields a density gradient within which DNA will migrate to form a band where its density equals that of the gradient, and the mol% G+C is directly related to this buoyant density. These methods have accuracies of about ±0.4 and ±1.0 mol% G+C respectively and both require a reference DNA of known mol% G+C, commonly that of *Escherichia coli*, for comparison. A recently developed, rapid method is unaffected by RNA contamination and is claimed to be of superior accuracy. The mol% G+C is calculated from ratios of nucleosides separated by HPLC after enzymatic degradation of the DNA.

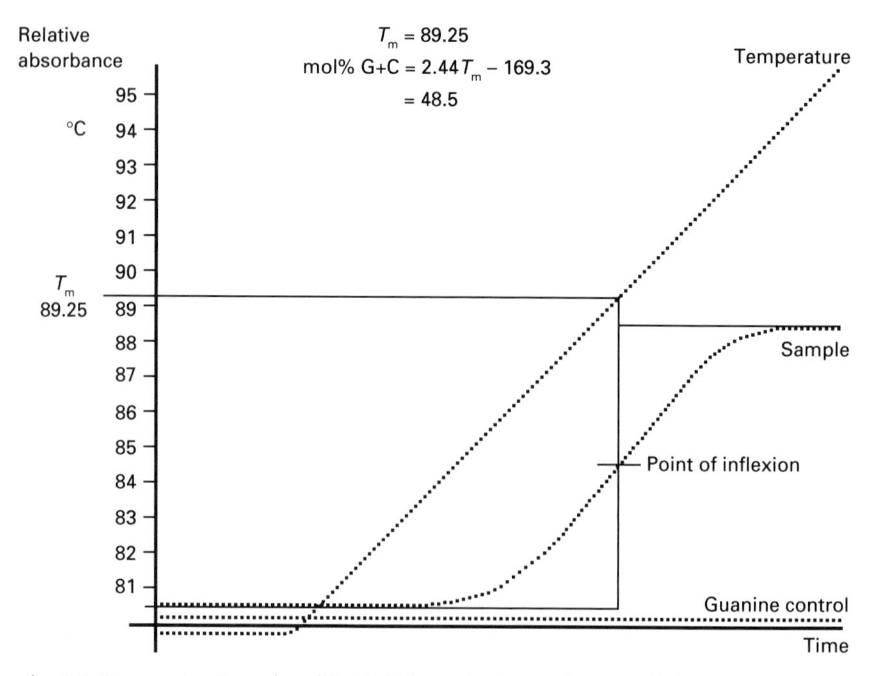

Fig. 3.2 Determination of mol% G+C from the hyperchromic shift accompanying thermal denaturation of DNA. The melting temperature (T_m) coincides with the point of inflexion (which is found midway between the top and bottom plateaux of the sample curve), and is determined from the temperature curve, reading from the scale superimposed on the relative absorbance scale on the *y*-axis. Mol% G+C is then calculated using a simple formula.

DNA relatedness

Because nucleotide base pairing is specific, the sequence of a DNA strand determines that of its complementary strand. At its T_m, as we have seen, or at a pH above 12, dsDNA dissociates (denatures) into ssDNA which will reassociate at a temperature 15–30°C below T_m, or at neutrality, to give the original molecule. It is this property of renaturation that is used in DNA **relatedness** (or **homology**) studies; if ssDNA from one organism is mixed in the appropriate conditions with ssDNA from another organism, hybrid (**heterologous**) molecules may form. The more genetically closely related the two organisms are the more nucleotide base sequences they will have in common, and so the greater the degree of hybrid formation (**hybridization**) that will occur. The amount of mismatch in the heterologous duplex molecule can be determined by comparing its T_m (called the $T_{m(e)}$) with that of a **homologous** duplex; the smaller the difference between their thermal stabilities ($\Delta T_{m(e)}$), the fewer the mismatches in the hybrid.

Whereas traditional characterization methods probably reflect no more than 10% of the genotype, DNA relatedness represents a considerably larger part of the genome. None the less, classifications based upon the results of the two approaches usually show general agreement. In the study of groups where few phenotypic tests are available for distinguishing between species, DNA relatedness is particularly useful. A further advantage of the method is that it brings us closer to a good definition of species: a widely used working-definition of **genomic species** (or **genetic species**) is a group of strains which show relatedness of 70% or more in optimal hybridization conditions, and 5°C or less $\Delta T_{m(e)}$. There are no such satisfactory definitions of genus or family; for the former 40–60% relatedness has been suggested but its rigorous application would seriously destabilize existing classifications whose taxa have been defined by widely varying criteria.

DNA reassociation experiments are very laborious and so only a limited number of reference and test strains, which must be carefully selected to represent the perceived diversity of the group under study, can be characterized. A full data set, amenable to numerical taxonomic analysis, may entail a prohibitive amount of work. A further limitation is that the data are always based upon comparisons with reference strains rather than an independent data set being obtained for each strain.

Determination. Many procedures have been developed for measuring DNA sequence similarity. All but one of the five methods commonly used require radioactive reference DNA, usually labelled with ^{14}C, and are of two types: immobilized DNA and free solution reassociation. In the first type of assay DNA is usually immobilized on nitrocellulose filters and relatedness is determined by **direct binding** or **competition** experiments. In the former (Fig. 3.3) ss test DNA is heat-fixed to a membrane which is then treated to prevent any further non-specific binding. It is then incubated with ss radiolabelled reference DNA and, after washing, the radioactivity bound is measured. Relatedness is calculated as a percentage of the counts

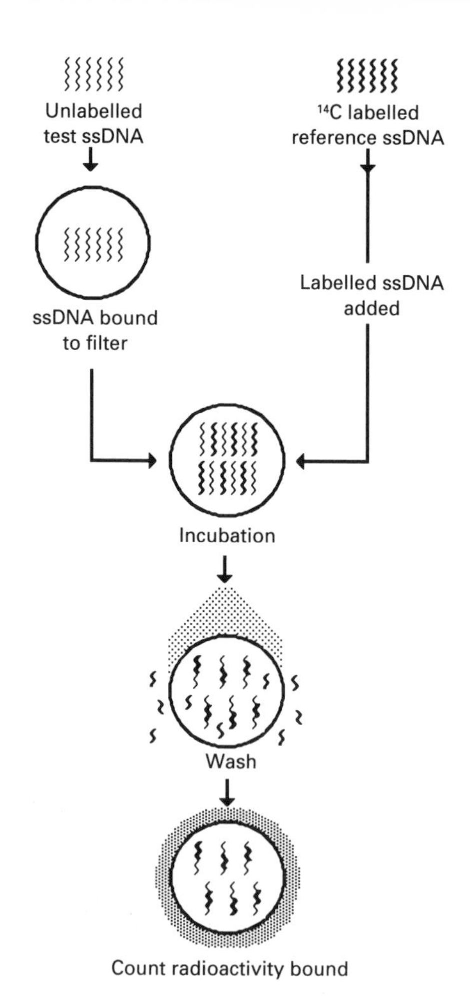

Fig. 3.3 Immobilized DNA reassociation assay.

bound when unlabelled reference DNA is fixed to the filter. The problem of getting consistent amounts of DNA onto the filters is overcome by the competition method (Fig. 3.4) in which unlabelled *reference* ssDNA is fixed to the membranes instead, and labelled reference ssDNA competes with an excess of unlabelled ss test DNA for the binding sites. Thus, the lower the counts bound the better the test DNA has competed, and so the higher its relatedness with the reference DNA.

Reassociation in free solution is determined optically or by radiolabelled probe DNA. In the first method the rates of reassociation of test with reference DNA and of each DNA separately are monitored by falls in absorbances at 260 nm in an ultraviolet spectrophotometer. The closer the rate of heterologous reassociation is to the rates of homologous reassociation, the higher the relatedness (Fig. 3.5). In the other methods ss radiolabelled reference (**probe**) DNA is incubated with ss test DNA in excess (to prevent the probe reassociating with itself); the higher the relatedness the closer the rate of probe–test reassociation to the rate of test–test reassociation and so the higher the radioactivity of the duplexes formed. After incubation the remaining ssDNA is removed by hydroxylapatite

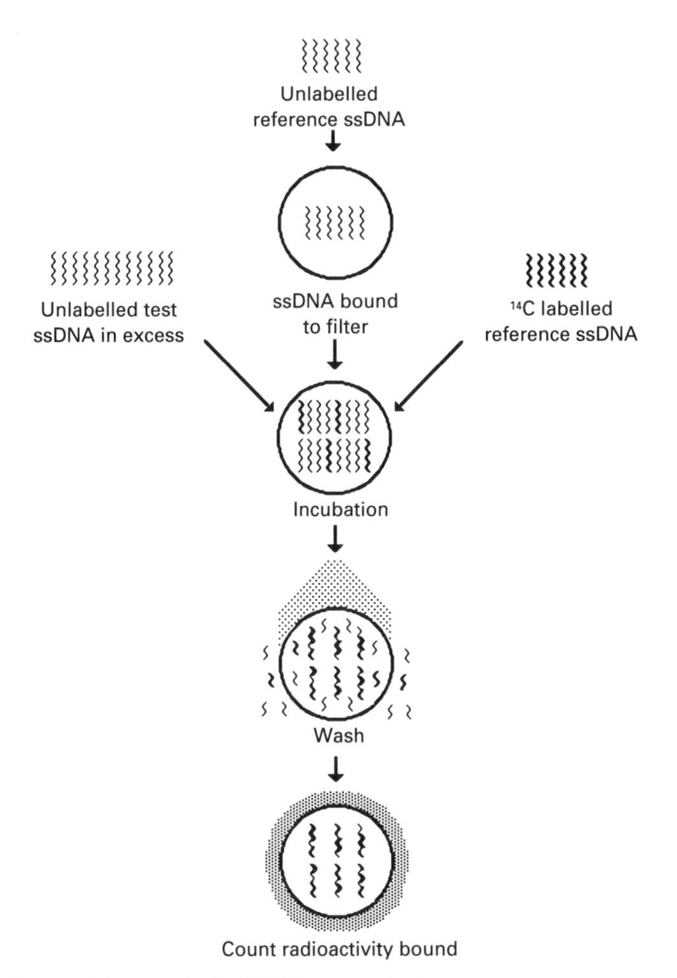

Fig. 3.4 Competition method of DNA reassociation assay.

chromatography or S1 nuclease: dsDNA binds to hydroxylapatite at low phosphate concentration, the ssDNA can be washed away and the dsDNA then eluted by raising the phosphate concentration; S1 nuclease hydrolyses ss- but not dsDNA and the latter can be recovered by acid precipitation.

A recently developed rapid method immobilizes ssDNAs in microdilution plate wells and hybridizes with photobiotin-labelled ssDNA. After hybridization and washing, the biotinylated DNA is quantitatively determined by binding a streptavidin-conjugated enzyme to the label and measuring with a fluorogenic enzyme substrate.

Reproducibility. Standardization of experimental conditions is most important; DNA purity, fragment size and concentration, ionic concentration, incubation period and temperature affect rates, levels and specificities of reassociation. At high salt concentrations, at temperatures lower than the optimum of 25–30°C below T_m, and with small DNA fragments for example, non-specific binding becomes a problem. Most parameters are fixed by the procedure adopted, but for closely related organisms the temperature may be raised to only 15°C below T_m to give more stringent

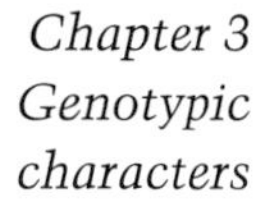

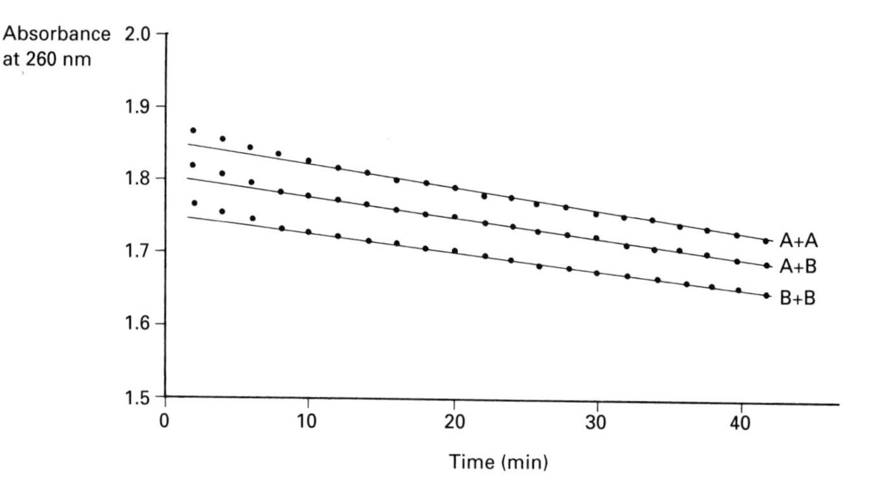

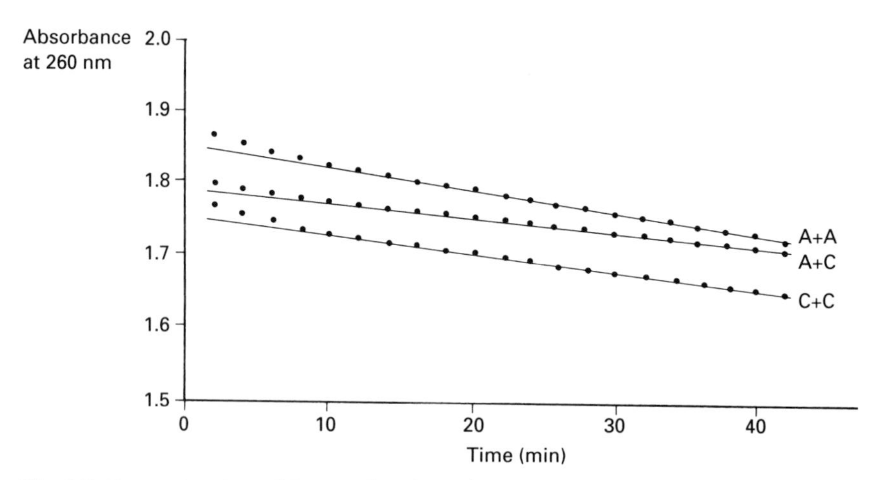

Fig. 3.5 Determination of DNA relatedness by reassociation rates in free solution. The rate for the heterologous reassociation A+B is similar to the rates for the homologous reassociations A+A and B+B, indicating that A and B are very closely related. The curve for the heterologous reassociation A+C shows a slower fall in absorbance, indicating that A and C are not closely related.

conditions. Given careful standardization, reproducibility is good and despite the variety of procedures available and the many variables involved, levels of agreement between different methods and studies are generally high.

Restriction endonuclease analysis

This technique, which is also called **restriction fragment length polymorphism analysis** or **genetic fingerprinting**, is simple and rapid, and requires very little DNA. It is of value in several kinds of taxonomic study, ranging from phylogenetic to epidemiological investigations.

The DNA is cleaved with various restriction endonucleases which break the molecule into fragments at different specific sites. The number and location of these sites are unique for each genome, and so the pattern of groups of fragments of different size is also unique for each. These fragment classes, which may number 50 or more, form the specific restriction pattern or fingerprint of an individual organism. Different endonucleases will give different patterns whose taxonomic values will vary with the group concerned.

A small amount of DNA, perhaps from a single plate or following PCR amplification, is extracted and purified. Non-specific fragmentation occurring at this stage does not appear to affect results, but modifications of bases (owing to methylation for example) may affect results, by reducing enzyme access to some sites. Portions of DNA are digested with a selection of restriction endonucleases and the fragments are separated by gel electrophoresis. The bands are stained with ethidium bromide and photographed on an ultraviolet transilluminator, and the negative film is scanned by a densitometer; the data may then be subjected to computer taxonomy.

For phylogenetic studies 16S rDNA (the genes for 16S rRNA) can be fingerprinted after amplification using PCR (Fig. 3.6). Primer target sites are not difficult to select, given the wealth of information available about conserved and variable regions of rRNAs.

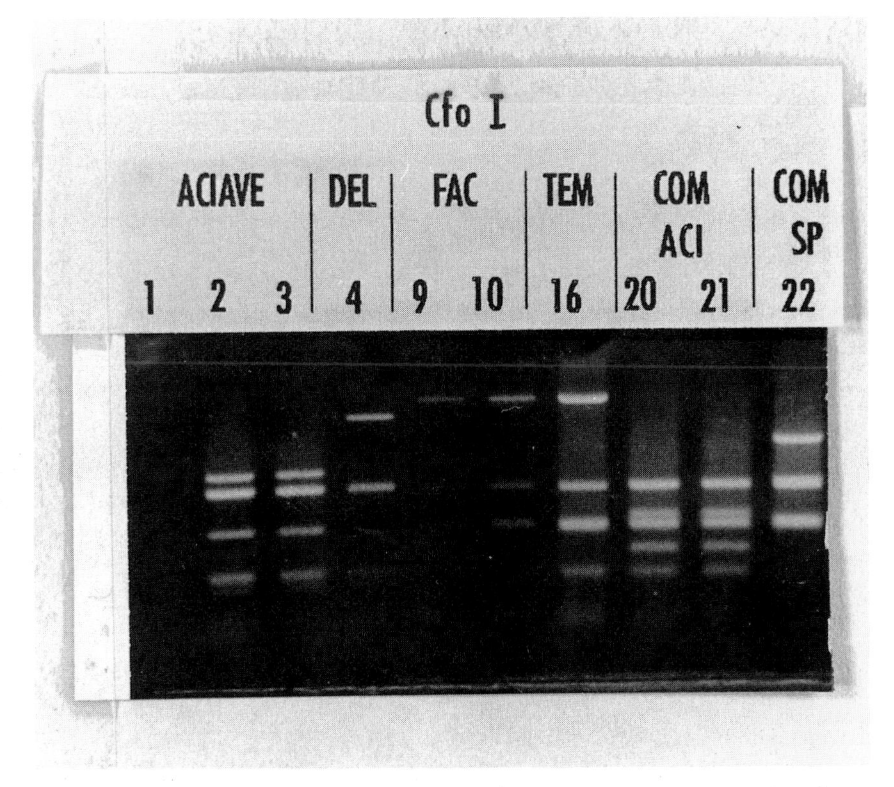

Fig. 3.6 PCR fingerprinting of *Acidovorax* and *Comamonas* species, cutting the PCR amplificate with restriction enzyme *Cfo*I. This shows discrimination between *A. avenae* (lanes 2 and 3), *A. delafieldii* (lane 4), *A. facilis*/*A. temperans* (lanes 9, 10 and 16, which are not distinguishable from each other), *C. acidovorans* (lanes 20 and 21) and *Comamonas* sp. (lane 22). Discrimination between *A. facilis* and *A. temperans* can be achieved using another restriction enzyme.

Ribosomal RNA analysis

Ribosomal RNA represents only a small part (about 0.3–0.4%) of the genome and the cistrons coding for it are highly conserved — they have evolved less rapidly than the rest of the chromosome — presumably because of the fundamental role of the ribosome. This feature of the rRNA molecule, the widely varying rates at which different parts of its sequence change, and the ability of these parts to change independently of each other, make it a valuable 'molecular clock'; very distantly related taxa have the most highly conserved sequences in common and yet the faster-evolving sequences are of value in comparing more closely related organisms. Ribosomal RNA sequence data have revolutionized bacterial systematics not only by providing information useful at generic and higher levels, but also, more excitingly, by giving insights into bacterial evolution. Furthermore, because rRNA is present in all organisms (viruses not being classed as organisms) the same approach is applicable to eukaryotes.

The three rRNAs are classified by their sedimentation rates as 23S, 16S and 5S, which have chain lengths of about 3300, 1650 and 120 nucleotides respectively, and can be separated by gradient centrifugation or electrophoresis. Until recently direct and complete sequencing of the larger rRNA molecules was not feasible as a routine approach; instead, sequence data were sought indirectly by DNA–rRNA hybridization or partial sequences were obtained semi-directly by oligonucleotide cataloguing. For prokaryotes, complete sequencing of 5S rRNAs waned in favour of 16S rRNA cataloguing for various reasons, including the greater information content of the larger molecule, and now essentially complete sequencing of the latter is routine. Whether 23S rRNA sequencing would be rewarded with much further useful information is unknown. Cataloguing and 5S rRNA sequencing enabled the tentative construction of phylogenetic trees and the recognition of major bacterial divisions. The development of direct and rapid sequencing of 16S rRNA has led to great improvements in the resolution of classifications based upon this molecule, and the great value of the method, especially as part of a polyphasic study, will become clear from later chapters. There is a danger in using this approach alone at the species level as its resolving power may be limited by experimental error, and because phenotypically similar organisms with very high 16S rRNA sequence similarities can show low DNA relatedness.

DNA–rRNA hybridization

Because only one DNA strand is used as a template for RNA synthesis, and RNA molecules are single stranded and do not pair with each other, rRNA relatedness is determined by hybridization with complementary ssDNA. The commonest approach hybridizes ^{14}C-labelled 16S or 23S RNA with ssDNA immobilized on nitrocellulose filters at temperatures 25–35°C below T_m. After washing and RNase treatment the amount of rRNA (i.e. radioactivity) bound is measured. The filters are then subjected to a series of temperature increases in 5°C steps up to 95°C, and radioactivity eluted at each step as the hybrids are denatured is measured to give values for thermal stability ($T_{m(e)}$). Problems can include renaturation of the rRNA, owing to its stable secondary structure, and difficulty in calculating the hybridization yield. An alternative method uses direct hybridization of rRNA fragments obtained by alkaline cleavage, and calculates yields by correcting for different genome lengths and operon numbers. Other approaches include competition and free solution methods, and in all cases purity of reagents is essential and must be assayed for. Hybridization is relatively simple and rapid, but it is reliant on reference strains, and is not as accurate as sequencing methods. None the less, the use of the $T_{m(e)}$ method at the University of Gent has led to major improvements in the classification of the *Proteobacteria*.

Oligonucleotide cataloguing

Prior to the development of direct sequencing methods, a large amount of taxonomic information was extracted from 16S rRNA by sequencing short lengths (**oligonucleotides**) rather than the whole molecule. Treatment of rRNA with T1 ribonuclease yields perhaps 600 oligonucleotides with lengths of up to 20 bases, each with a guanine residue at the 3' end. These were labelled *in vitro* with ^{32}P at the 5' end, separated by electrophoresis and TLC, and visualized by autoradiography. Oligonucleotide sizes and compositions determined the positions of the spots, which were removed, digested with alkali and then sequenced by two-dimensional TLC. Sequences were tabulated to form a catalogue for each strain and these were compared by numerical methods. The method was expensive and slow, taking 4–5 weeks to obtain a single catalogue, and reliable interpretation of two-dimensional TLC data required considerable experience.

Direct sequencing

Methods for DNA sequencing have been adapted for application to rRNAs. All possible oligonucleotides are obtained by digestion with various endonucleases and then radioactively labelled. Sequences are read from electrophoresis gels of fragments differing from each other by a single residue. In rapid, direct sequencing of 16S rRNA the molecule does not have to be isolated. Instead, it is targeted by synthetic oligodeoxynucleotide primers which are complementary to the universally conserved 16S rRNA sequences. Reverse transcriptase then generates copy DNA for sequencing. The procedure takes only a few days and full sequence data are compatible with oligonucleotide catalogues. The sequence data are aligned and relatedness values determined; nucleotide substitution rates (K_{nuc}) or 'evolutionary distance values' are then calculated and commonly subjected to distance matrix or maximum parsimony analysis (see Chapter 4) to produce phylogenetic trees.

All sequencing work requires highly purified molecular fragments and, especially as the procedures are complicated, strict standardization and cleanliness are essential at all stages to avoid erroneous results; great care must be taken to minimize errors when reading the gels. Problems of interpretation are considered in Chapter 6.

Gene transfer

Horizontal transfer (loosely called **recombination**) of chromosomal genes is a potentially useful tool for investigating relationships among bacteria, but although studies of **transformation**, the environmental transfer of free DNA, have given valuable insights for several taxa, such approaches have been little used. It has been suggested that statistical analyses of gene sequences from closely related organisms, to determine the likelihood of recombination occurring, can allow the biological species definition (i.e. interbreeding) to be applied to bacteria.

Taxonomists have concerned themselves more with the effects of extra-chromosomal elements on classification and identification. A few unstable plasmid-borne characters are unlikely to have a major effect on a polythetic classification but in identification schemes however, it may be important to avoid them.

4 Similarity and arrangement

The different kinds of bacteria are not separated by sharp divisions but by slight and subtle differences in characters so that they seem to blend into each other and resemble a spectrum.

S.T. Cowan & K.J. Steel, 1965, *Manual for the Identification of Medical Bacteria*

Introduction

Having characterized their strains, taxonomists wish to discover what groups are implied by the data they have gathered. To reveal these **taxonomic structures** the data for each OTU must be objectively compared with those for every other OTU. The large amounts of information generated by characterization studies cannot be analysed by subjective intuition, which would waste such carefully collected data, but must instead be processed by computer using methods that are in many cases beyond the capabilities of the human mind. It should be appreciated that numerical taxonomy now embraces genomic as well as phenotypic characters.

Recording and coding data

The form in which the computer program will accept the data should be determined at the beginning of the study; the less the results have to be manipulated the smaller the chance of error. It is therefore best to prepare scoring sheets for the direct recording of raw data at the bench, which have corresponding data sheets in which the computer-ready codes are entered. The resulting table of strains versus characters is called the **data matrix**. The matrix must be complete as any omissions would prevent computation, but rather than any missing value necessitating the exclusion of an OTU and a character, most programs have a **no comparison** (NC) facility whereby gaps are indicated with a special symbol such as an asterisk and ignored during analysis; this is valuable if some results are inherently unobtainable for some strains, e.g. spore shape and position in non-sporeformers.

Characters may be binary, qualitative multistate, or quantitative. **Binary** characters have two states — presence/absence (of spores, for example) or positive/negative (as for the catalase test) — and are commonly scored as 1, 0 respectively. **Qualitative multistate** characters are those that exist in more than two mutually exclusive states but cannot be measured quantitatively, e.g. colony colours. They may be regarded as several individual binary characters, with a strain producing violet colonies scoring NC for yellow colonies for example, or one character with several states, such as 1, 2, 3, etc., the difference between each pair being the same. **Quantitative** characters may be scored directly or as a small number of clearly distinguishable classes (ordered multistate); examples are antibiotic disk inhibition zone radii measured in mm, or as classes 0, 1, 2 and 3 representing 0 mm (i.e. resistance), ≤5 mm, ≤10 mm and ≤15 mm respectively. A few programs require all data to be in binary form so that quantitative characters have to be split into several individual ones and coded additively:

	Character		
	1	*2*	*3*
0 mm (resistant)	0	*	*
≤5 mm (slightly sensitive)	1	0	*
≤10 mm (moderately sensitive)	1	1	0
≤15 mm (very sensitive)	1	1	1

The asterisks indicate redundant entries which are scored NC.

Scaling

Characters are given equal weight initially but some quantitative ones having large values may exert excessive weight with some methods of estimating similarities and so require scaling. The simplest approach is **ranging**, where the smallest observed value for a character is subtracted from all values and the results are divided by the range:

$$\text{ranged value} = \frac{(\text{value} - \text{lowest value})}{(\text{highest value} - \text{lowest value})}$$

Thus, the set of values 1 mm, 5 mm, 9 mm and 40 mm for colony diameter becomes 0, 0.1, 0.2 and 1.

The coded data matrix is then entered into the computer and checked for errors (verification). If the program cannot handle all the data at once, the constraint usually being in the number of OTUs, several computer runs are made. A final run containing representatives from each group discovered links with the previous runs.

Classification methods

Bacteriologists use two different kinds of approach to classification. In **hierarchical** methods groups of similar OTUs are recognized and assigned, without overlapping, to a succession of ranks of different seniority (a hierarchy). The arrangement is most often depicted as a **dendrogram** (from the Greek *dendron*, tree) as used for family trees (Fig. 4.1a). In **non-hierarchical** methods the relationships between OTUs are represented in terms of distance in space and displayed as maps in two or three dimensions (Fig. 4.1b); groups are not formed and have to be recognized by eye.

Hierarchical methods have been most widely used for bacterial classification because in traditional biological taxonomy the Linnaean system requires distinct ordered classes. However, classes are artificial concepts and, especially among the bacteria, spectra of strains or species complexes

a.

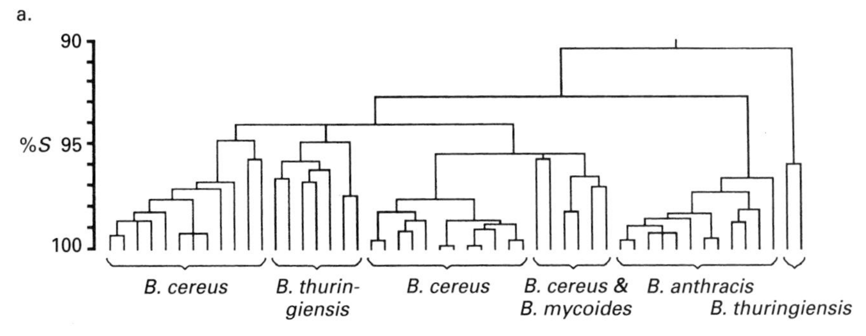

b.

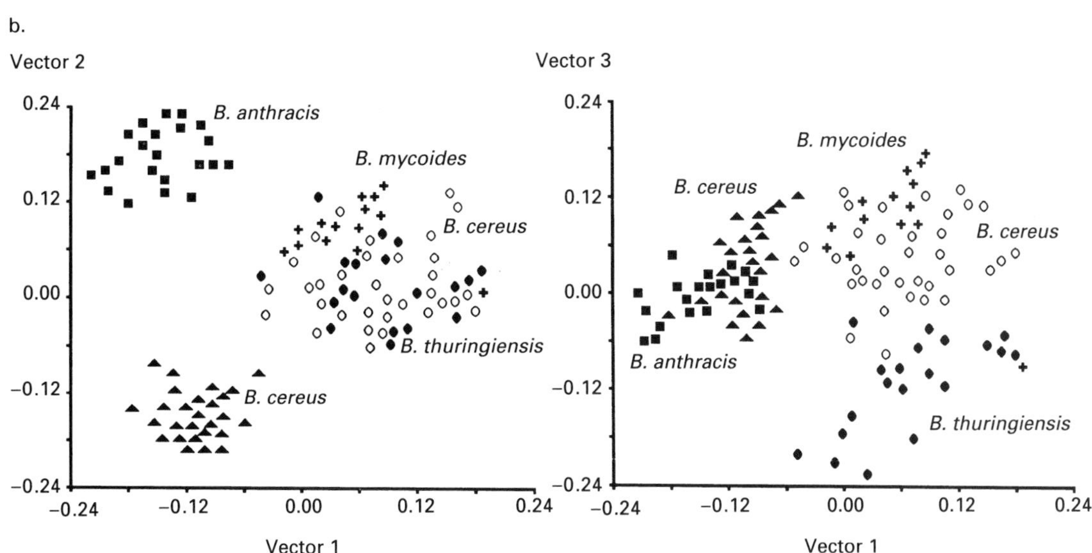

Fig. 4.1 a. Dendrogram. b. Ordination plots.

are frequently encountered; such data may be more amenable to non-hierarchical analysis. Classifications are readily produced by hierarchical methods and these represent relationships between close neighbours well, but information on relatedness between more distant groups tends to be lost and their relationships become distorted. While dendrograms have the advantages of conveniently summarizing hierarchical classifications, such two-dimensional representations inevitably entail further information loss.

Non-hierarchical techniques such as ordination, while not always representing distances between close neighbours faithfully, do not distort distances between major groups and may reveal relationships, such as those within spectra, that may not be apparent in hierarchical analyses. The data may, however, be too complex for satisfactory display in a two- or three-dimensional plot. Thus, hierarchical and non-hierarchical approaches may be complementary, and should be used together as both are commonly available in computer programs for numerical taxonomy.

Similarity and dissimilarity

Classifications are based upon correlations among characters that give measures of **resemblance** between the OTUs. Therefore, the first step in analysis is to construct a matrix of **similarities** or dissimilarities (**distances**) between every pair of OTUs from the data matrix. The range of methods for calculating resemblance is large but only a few such coefficients are commonly used in bacteriology.

Similarity coefficients

The simplest coefficients are those for binary data. For two OTUs there are four possible combinations of results in any one test:

		OTU A	
		+	–
OTU B	+	α	β
	–	γ	δ

i.e., both positive (α), only one positive (β and γ), and both negative (δ).

The **simple matching coefficient** (S_{SM}) is calculated as the number of data matches between two OTUs divided by the number of characters:

$$S_{SM} = \frac{\alpha + \delta}{\alpha + \beta + \gamma + \delta}$$

If for two OTUs and 100 tests the results are α, 50; β, 20; γ, 10; and δ, 20, the similarity (S) between them is:

$$\frac{50 + 20}{50 + 20 + 10 + 20} = 0.7 \text{ or } 70\%$$

In S_{SM} the data are treated symmetrically with negative matches having the same weight as positive ones. If characters inapplicable to both OTUs outnumber those that are just negative for them, then S_{SM} may indicate excess similarity. Furthermore, in bacteriology a positive test result is often considered to be more informative than a negative one, especially with slow-growing or weakly-reacting organisms, and an asymmetric coefficient such as **Jaccard's** (S_J), in which negative matches are ignored, may be used:

$$S_J = \frac{\alpha}{\alpha + \beta + \gamma}$$

with the data used above this gives a similarity of 0.625 or 62.5%.

Groups based upon S_J join at lower similarities and tend to be less compact and clearly defined than those based upon S_{SM} but, in practice, the results are usually comparable; if they are not it indicates that negative matches are having a strong influence — as they might with unreactive organisms. In addition to the considerations noted, choice of coefficient may depend on whether the taxonomist is a lumper or a splitter.

Gower's general similarity coefficient (S_G) allows mixed binary, qualitative multistate and quantitative characters to be used. The weighted averages of all similarity values between pairs are calculated from their character scores:

$$S_G = \frac{\text{sum of weighted scores}}{\text{number of characters}}$$

All invariant characters are excluded and NCs are given zero weight. With binary characters positive matches are given a weight of 1 and score 1; the coefficient is equivalent to S_J or S_{SM} acccording to whether the weight for negative matches is set to 0 or 1. For quantitative characters the score is calculated by scaling each character on its range as described above:

$$\text{score} = 1 - \frac{(|\text{value for OTU A} - \text{value for OTU B}|)}{\text{range}}$$

for the range 0–5, the score for the results 0 and 4 would be the same as that for 1 and 5 (0.2 or 20% similarity) whereas between 0 and 5 it would be zero, and this may be unacceptable if negative reactions are due to less vigorous growth rather than the absence of the characters. This is circumvented by setting the range higher than that observed; with a range of 6 the results become 0.33 (or 33%) and 0.16 (16%) respectively, but identical states still score 1.

Distance coefficients

Another coefficient for quantitative data is **taxonomic distance** (d). The OTUs are viewed as points in a multidimensional space (**hyperspace**), with one dimension for each character, and resemblance is expressed as **Euclidean distance**. This is the shortest distance between two points whose positions are defined by axes at right angles to each other (the distance XZ in Fig. 4.2):

$$d = \sqrt{(\text{OTU A}a - \text{OTU B}a)^2 + (\text{OTU A}b - \text{OTU B}b)^2 \ldots (\text{OTU A}n - \text{OTU B}n)^2}$$

for n characters. On the basis of the data for the two characters used in Fig. 4.2 the taxonomic distance between OTU A and OTU B is 4.12. We can visualize a third axis, at right angles to the other two, but further axes can only be defined mathematically. The measure increases with the number of characters and is not a percentage, but ranges from zero (= 100% similarity) to infinity. Dividing by the number of characters gives the average distance (2.06 for the example). For binary data $d = \sqrt{1 - S_{SM}}$ and is a measure of mismatches (β and γ above).

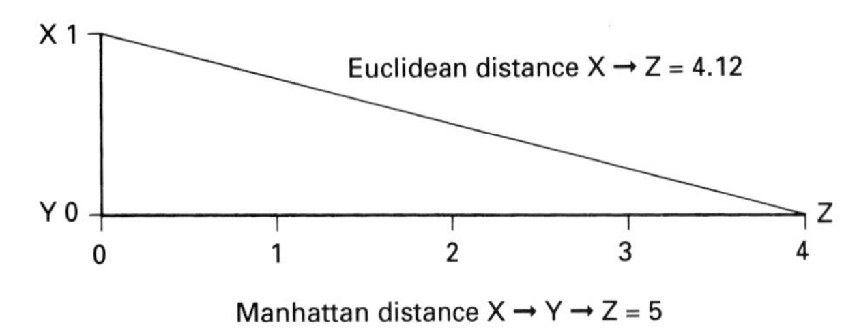

Fig. 4.2 Distance measurements.

When using standardized tests for characterizing a diverse range of bacteria which vary in their requirements for optimal growth, some dissimilarities between strains may be solely caused by differences in their growth rates. Sneath divided total dissimilarity ($D_T = 1 - S_{SM}$; a **city block** or **Manhattan distance** such as XYZ in Fig. 4.2) between two strains into components of **vigour** (D_V) which represents differences in growth rates or times of reading tests, and **pattern** (D_P) which represents differences between equally vigorous strains and is taxonomically the more interesting. For binary data:

$$D_V = \frac{\beta - \gamma}{\alpha + \beta + \gamma + \delta} \quad \text{and} \quad D_P = \frac{2\sqrt{(\beta\gamma)}}{\alpha + \beta + \gamma + \delta}$$

(for a Jaccard type of measure δ would be omitted). Also, $D_T^2 = D_V^2 + D_P^2$ and $D_T = d^2$. With the 100 character set used earlier $D_T = 0.3$, $D_V = 0.1$ and $D_P = 0.283$. The complement of D_P is **pattern similarity** ($S_P = 1 - D_P$) which measures the similarity between two OTUs when the vigour component is omitted (0.717 or 71.7% for the sample data). Such coefficients are only appropriate when the characters are based on metabolic activities and logically scorable as positive or negative (not, for example, colony shape), but allow useful corrections, and checks on the contribution of negative matches to the similarities of unreactive organisms.

The unsorted resemblance (or similarity) matrix generated by one of the above coefficients is a table of OTUs vs. OTUs in their original order and the entries are resemblance estimates for each pair of OTUs. Because the resemblance between OTU A and OTU B is the same as that between OTU B and OTU A, only one triangular half of a matrix is printed; also, the self comparisons along the main diagonal with similarities of 1 (or 100%) or distances of 0 are usually omitted (Fig. 4.3a).

Taxonomic structure

An unsorted resemblance matrix cannot be extensively interpreted without recourse to computer analysis by hierarchical or non-hierarchical methods, preferably both, which reveal the taxonomic structure by reducing patterns in phenetic hyperspace to just a few comprehensible dimensions.

Hierarchical methods

Hierarchical methods usually seek the two closest OTUs, which then form a compact group or **cluster** to which other OTUs (or independently formed

Fig. 4.3 (opposite) a. A similarity matrix for six strains. b. The progress of cluster analysis by three methods; only the changes in cluster membership are shown. c. Phenograms derived from the three cluster analyses. d. Ordered similarity matrix. e. Shaded similarity matrix. f. A minimum-spanning tree drawn in perspective.

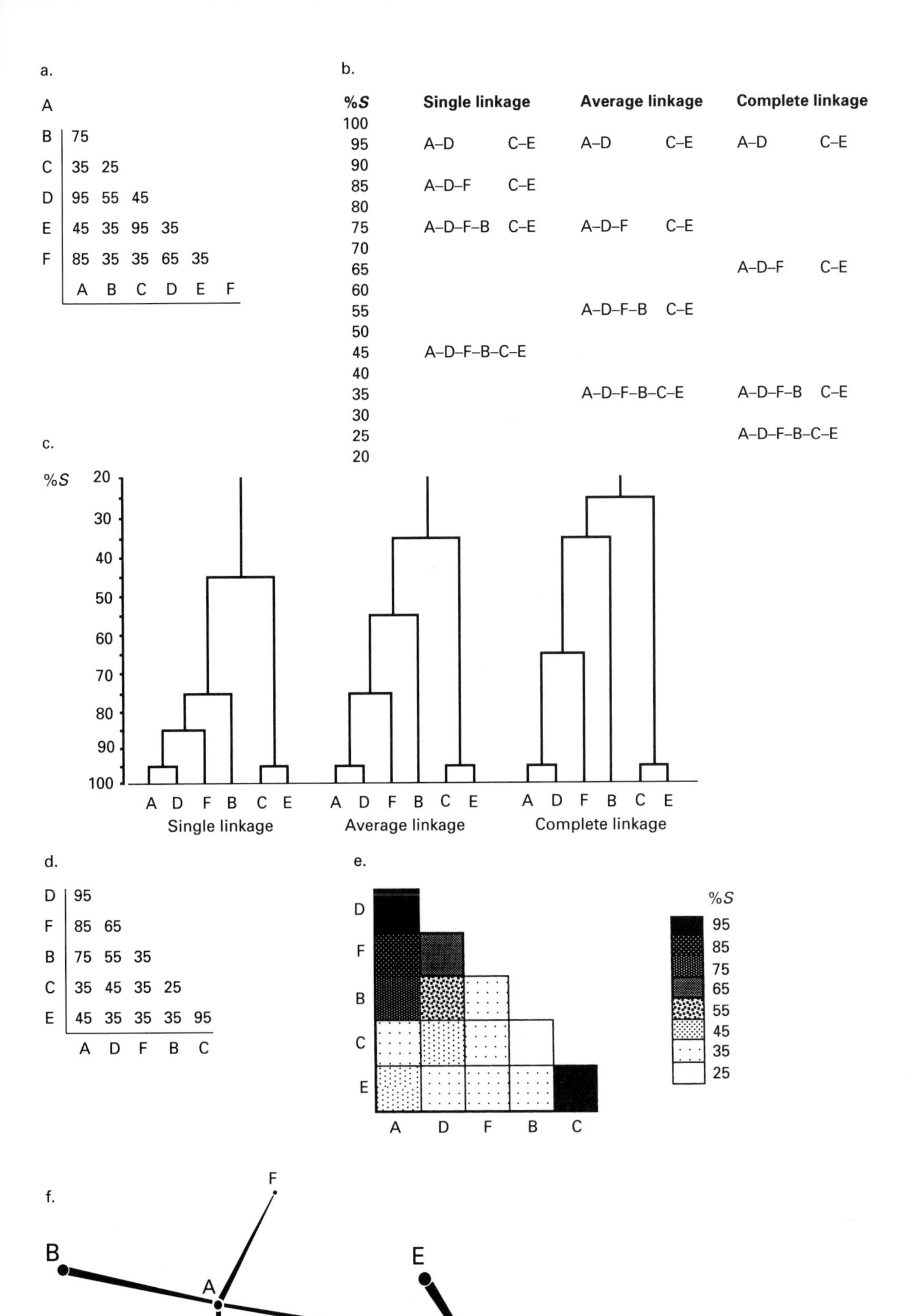
a.
A
B 75
C 35 25
D 95 55 45
E 45 35 95 35
F 85 35 35 65 35
A B C D E F
b.
%S
Single linkage
Average linkage
Complete linkage
100
95 A–D C–E A–D C–E A–D C–E
90
85 A–D–F C–E
80
75 A–D–F–B C–E A–D–F C–E
70
65 A–D–F C–E
60
55 A–D–F–B C–E
50
45 A–D–F–B–C–E
40
35 A–D–F–B–C–E A–D–F–B C–E
30
25 A–D–F–B–C–E
20
c.
%S
20
30
40
50
60
70
80
90
100
A D F B C E
Single linkage
A D F B C E
Average linkage
A D F B C E
Complete linkage
d.
D 95
F 85 65
B 75 55 35
C 35 45 35 25
E 45 35 35 35 95
A D F B C
e.
D
F
B
C
E
A D F B C
%S
95
85
75
65
55
45
35
25
f.
F
B
A
E
C
D

clusters) showing high resemblance may join according to the criteria of the **cluster analysis** method chosen. The procedures for three commonly used methods are illustrated by Fig. 4.3b: the similarity matrix in Fig. 4.3a is repeatedly scanned at decreasing similarity levels, a decrement size of 5% having been chosen (see Error and Goodness of Structure below), and the two clusters AD and CE are found at the 95% level. Then, by the **single linkage** (or nearest neighbour) method, F joins the cluster AD at the 85% level, as this is the highest similarity between it and any one member of the cluster (in this case A). However, F and D show only 65% *S*, so the cluster ADF appears more compact than the data imply and taxonomic space has been contracted. With the **average linkage** method (or unweighted pair group method using arithmetic averages, UPGMA) the candidate OTU joins a cluster at the average similarity between it and the existing cluster members, and taxonomic space is conserved; thus F joins AD at 75% *S*. The analyses continue as shown in Fig. 4.3b until all OTUs have been clustered. In these examples only the S levels at which clusters join differ, but in larger studies the order of joining, and so the taxonomic structure implied, may also differ.

The clustering method selected can therefore influence the classification profoundly, yet choice tends to be somewhat subjective and guided by what the taxonomist hopes to see. Space-dilating methods such as **complete linkage** (the antithesis of single linkage) may give clearly defined clusters at the expense of faithfulness to the similarity matrix, whereas space-contracting methods may produce poorly separated clusters that merge too early and give a chaining effect. Opinions differ as to which method is best; however, UPGMA distorts the parent similarity matrix the least and other approaches may be used to complement it.

Of the several different ways for displaying the results of cluster analysis, **phenograms** (dendrograms based on phenetic data) are the most widely used. Those derived from Fig. 4.3b are shown in Fig. 4.3c. It must be appreciated that relationships are only measured by vertical branch lengths; the order ADFBCE is somewhat arbitrary and could be changed to DAFBEC without altering the structure implied (the phenogram can be rotated like a mobile). Large phenograms may be simplified by representing compact clusters as triangles whose base widths are proportional to the numbers of OTUs they contain (Fig. 4.5a). As phenograms distort relationships between major groups they may be supplemented with **ordered similarity matrices** (Fig. 4.3d) in which interpretation has been simplified by shading according to the *S* level; the example in Fig. 4.3e indicates that cluster CE has higher similarity to AD than it does to B and F, and this was not apparent from the phenograms. Shaded matrices consume a lot of space however, and the information is often conveyed better by ordination plots. A single linkage cluster analysis may also be presented as a **minimum spanning tree** in which all the OTUs are coupled by a series of links so as to give a tree of minimum length overall (Fig. 4.3f).

Analysis of genotypic data for phylogeny

Phylogenies are usually presented either as unrooted trees or **cladograms** as these best summarize the relationships inferred, but none of the several quite different methods currently used for their construction is entirely satisfactory, as they all make assumptions about the evolutionary process (see Chapter 6). Cladograms are rooted, directional trees (i.e. dendrograms) based upon phylogenetic data, and their roots or origins represent postulated common ancestors. Trees without roots may be less easy to understand, but are usually preferred to cladograms because they exist in fewer possible versions and so less computation is required.

DNA–rRNA Hybridization. Sequence similarities of DNA–rRNA hybrids are measured by their thermal stabilities expressed as $T_{m(e)}$ (see Chapter 3) which can be used rather like a similarity coefficient — but without cluster analysis since not all possible pairs are hybridized and a similarity matrix is not formed — to construct dendrograms, as the higher the thermal stability of the hybrid the closer the relationship between the organism supplying the probe rRNA and the source organism of the test DNA.

Sequence Data. The first step in comparing sequences is to align them empirically. The most obviously homologous regions are aligned initially and then gaps are introduced as necessary to allow for deletions and insertions so as to bring all of the matching sequences into register. Where sequences are quite diverse there may be some regions of ambiguous alignment, and these are omitted from subsequent analysis. A measure of evolutionary distance such as K_{nuc} (nucleotide substitution rate) is then derived; one such formula is:

$$K_{nuc} = -3/4 \ln[1 - 4/3(d/s + d)]$$

where K_{nuc} is evolutionary distance between two sequences, d is the number of alignment positions where they have different nucleotides, and s is the number of positions where nucleotides are the same. A simple 25% difference becomes 0.3041 or 30.41%. Further corrections may be made to allow for gaps resulting from additions and deletions. Trees are then constructed using distance matrix methods or maximum parsimony analysis.

Distance Matrix Methods. These are based upon the fraction of positions in which sequences differ and a distance is usually expressed as the average number of point mutations occurring per position. Evolutionary distances expressed as simple percentage differences will be underestimates for all but very closely related molecules because multiple and back-mutations have not been taken into account. Making allowance for these is difficult owing to variations of evolutionary rates within and between molecules, so the various formulae which attempt to correct for them can give only estimates of evolutionary distance.

The tree of shortest length that represents the distance matrix most faithfully (usually according to a least sum of squares analysis) is then sought. Even with moderate numbers of sequences the number of possible trees for testing will be unmanageably large, so approximation methods are employed.

Maximum Parsimony Analysis. Instead of summarizing relationships as average similarities or distances, this method examines the evolution of each sequence position independently and seeks the tree that requires the fewest mutational events. A kind of minimum spanning tree based upon Manhattan distances is constructed and branching points representing hypothetical ancestors are introduced where they serve to shorten the tree. The number of possible trees is again huge (nearly 35 million cladograms and over 2 million unrooted trees for only 10 OTUs), so methods that avoid examining all the possibilities are used. Like cluster analysis, maximum parsimony analysis tends to show rapidly evolving lines as branching too deeply and so distance matrix methods appear to be more reliable.

Cluster Analysis. Oligonucleotide catalogues were commonly compared by using a Jaccard type of similarity coefficient to give the ratio of matches to the sum of matches and mismatches:

$$S_{AB} = \frac{2N_{AB}}{N_A + N_B}$$

where N_{AB} is the number of bases in oligonucleotides of length greater than five which are common to the two catalogues A and B, and N_A and N_B are the total numbers of bases in oligonucleotides of length greater than five in the two catalogues A and B respectively. A cladogram could then be constructed from the similarity matrix by average linkage cluster analysis. Cluster analysis has also been used for the analysis of taxonomic distance matrices derived from sequence data, but while it is computationally the least demanding approach it is also the least accurate, being particularly sensitive to varying rates of evolution. Rapidly evolving lines branch off too early in trees based upon cluster analyses and this tendency can only be corrected for partially.

Non-hierarchical methods

The non-hierarchical techniques most often used in bacterial taxonomy are **principal components analysis** (PCA), **principal coordinates analysis** (PCO) and **non-metric multidimensional scaling** (MDS). All of these are **ordination** methods which essentially reduce the number of dimensions (one for each character) in the data so that the relationships implied by the whole character set can be summarized by a much smaller number of dimensions, preferably only two or three, for viewing as maps or models.

Principal Components Analysis. The procedure for PCA is illustrated geometrically as a two-dimensional case in Fig. 4.4. In Fig. 4.4a the scores of

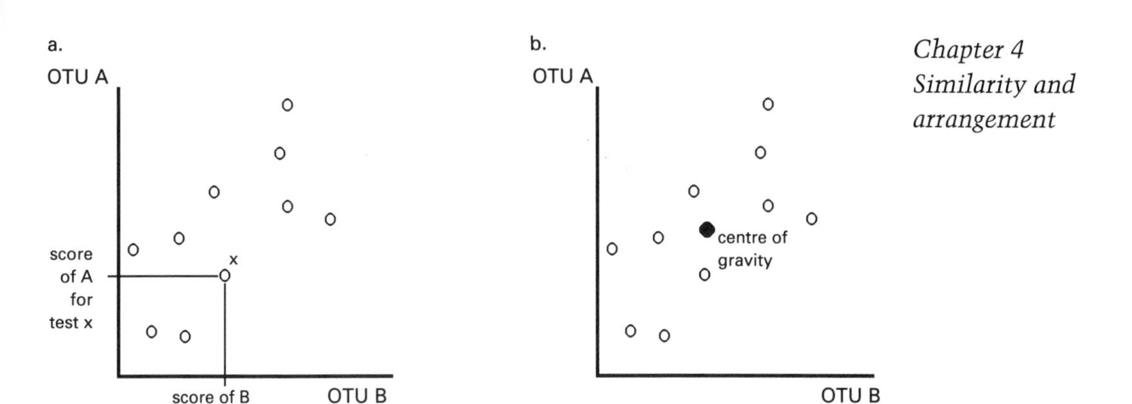

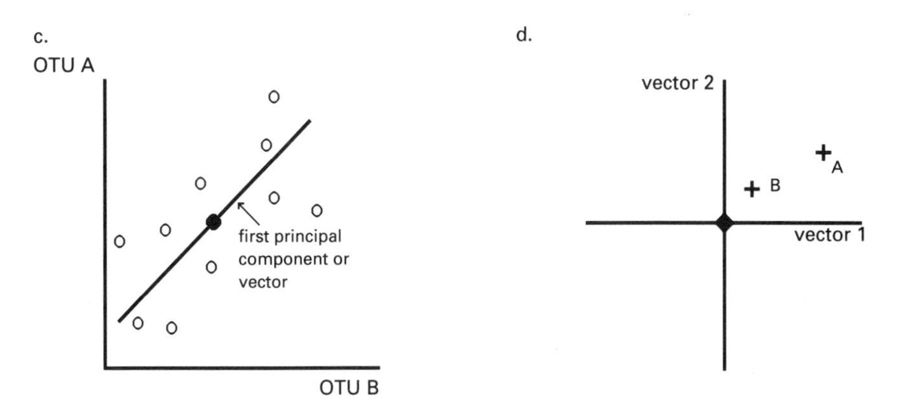

Fig. 4.4 The steps in principal components analysis for a two-dimensional case.

two OTUs for 10 characters are represented as a scatter plot. A correlation matrix of characters vs. characters is computed from the data matrix and the centre of gravity found (Fig. 4.4b). Axes are then rotated about this point so that the amount of scatter around the first axis is minimized (Fig. 4.4c). This first, and longest, new axis is the first principal component and it accounts for the largest possible portion of variance in the original data. An axis that accounts for the largest possible portion of the residual variance is then sought and becomes the second principal component, and so on until little or no variance is left. All the components are viewed as axes perpendicular to each other (**orthogonal**). The original character scores for each OTU may then be transformed to give a coordinate on each component axis. If the original data are correlated (that is, they imply a taxonomic structure) the first few components will collectively account for the majority of the total variance and can be used as axes for scatter plots of OTUs (Fig. 4.4d). None the less, PCA is primarily a character analysis, the results of which are especially valuable when seeking tests of high separation value for identification schemes.

A variant is **soft independent modelling of class analogy** (SIMCA) in which data from a group of OTUs are modelled separately by PCA; the plots preserve as much of the variance in the hyperspace as possible, and so allow for the existence of intermediate and outlying strains.

Principal Coordinates Analysis. Given the coordinates of a set of OTUs it is simple to calculate the Euclidean distances between each pair. Scaling methods such as PCO and MDS work the other way round and compute the coordinates of the OTUs from dissimilarity information; it is like drawing a map from a road atlas 'Distances Between Major Towns' matrix, but with multidimensional distances. PCO is closely related to PCA and if the distances used are Euclidean the results are the same. It is, however, an OTU analysis and the original data matrix is not required as the OTUs' coordinates along the principal components are computed from a distance or similarity matrix rather than a matrix of character correlations. Such an approach is potentially useful for the analysis of nucleic acid hybridization and immunological distance data.

Non-Metric Multidimensional Scaling. This technique generates new coordinates for the OTUs in a space of reduced dimensions so as to produce the most distortion-free, simplified, geometric model possible. It differs from PCO in attempting to preserve the rank order of the magnitudes of distances between OTUs and this results in a truer phenetic picture. An initial configuration is suggested, or formed, from a few randomly chosen OTUs, and the computer, working from a similarity or dissimilarity matrix, adds OTUs or adjusts their positions in small steps (**iterations**) as there is no direct way of finding the best solution. The results at each iteration are tested for distortion and the process continues until all OTUs are included and further adjustment fails to improve the representation.

Presentation. If the majority of variance in the data is accounted for by the first two axes, the taxonomic structure may be summarized by a two-dimensional scatter plot or map, and a minimum spanning tree may be used in combination with such a diagram to show which points are close in all dimensions. A series of two-dimensional plots may be used when a third axis is needed to separate overlapping groups, but three-dimensional models are preferable. These have frequently been presented as perspective drawings or 'ball and stick models' photographed from several angles, but three-dimensional computer projections may now be rotated and printed as required (Fig. 4.5b and c).

Error and goodness of structure

Before attempting to interpret the results the taxonomist should be confident that any errors inherent in the classification are not large enough to disturb the taxonomic structure seriously, but this is an area frequently neglected. Sources of error include: (i) inadequacies of representation in the culture collection; (ii) test inconsistencies; (iii) inaccuracies in scoring and recording characters; and (iv) distortions introduced by analysis. The first two of these should have been considered at the very beginning of the study and were discussed in Chapter 2. However, it is only at this later stage that the influence of test error (and, of course, errors in recording and handling data) can be estimated from the resemblance measures between duplicated strains.

Test error

In the similarity range of interest test errors may depress S_{SM} values by up to $2P$, and the effects are similar with other coefficients. Duplicated strains should show 100% S but rarely do, and if P is >10% overall (so that the duplicates show <80% S) a serious loss of taxonomic structure is implied. Most laboratories are able to keep P below 5%, but discrepancies between laboratories are often much greater and have important implications for identification schemes.

Confidence limits for resemblance coefficients

The decrements of resemblance level used in cluster analysis should be greater than one standard error (SE) for the branchings to be significant, and clusters may be regarded as significantly distinct if the lengths of the stems separating them in the phenogram are several times the SE; mean within- and between-cluster similarities give further useful information. The SE of the binomial distribution gives a satisfactory estimate of the confidence limits for given S_{SM} or S_J values if they are not extreme (see Sneath & Johnson, 1972):

$$\mathrm{SE} = \frac{\sqrt{(S)(1-S)}}{n}$$

for 0.7 (70%) S from 100 (n) characters the result is 4.6% which represents one standard deviation; applying 95% confidence limits from the normal distribution gives a range of 61% to 79% S.

Cophenetics and congruence

Distortion introduced by analysis can be estimated by comparing the values in the similarity matrix with those implied by the cluster analysis, or phenogram, to give the **cophenetic correlation coefficient**. Low correlation shows that the taxonomic structure is being distorted; if different approaches to analysis (such as using $D_{P,}$ or changing from single linkage to UPGMA) do not increase the correlation, it may be that hierarchical classification is inappropriate. Similarly, ordinations may be compared with their resemblance matrices.

Correlations of resemblance matrices may be determined to indicate congruence between classifications of a set of OTUs using different resemblance coefficients on one data matrix, or between those based upon different character sets such as phenotypic and genomic data. The latter approach is desirable in polyphasic studies, and it is also important to evaluate a numerical classification using independent taxonomic criteria such as DNA relatedness.

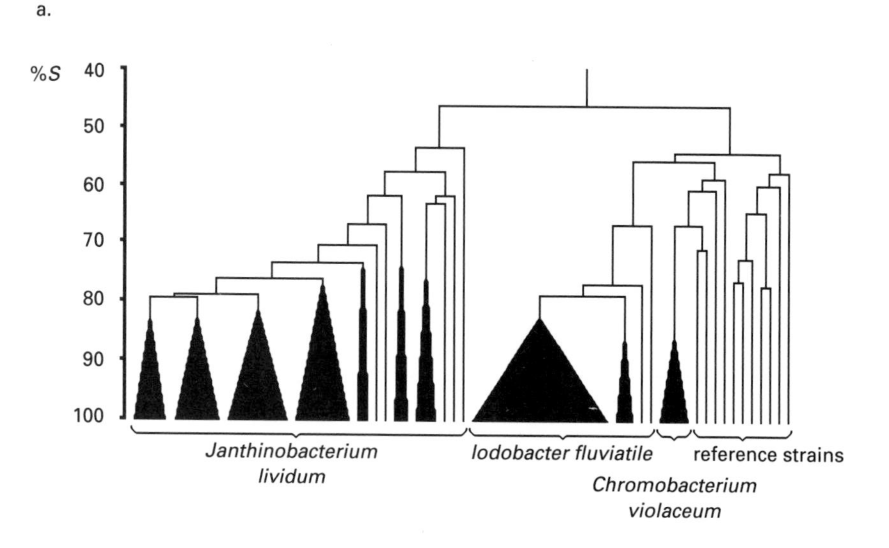

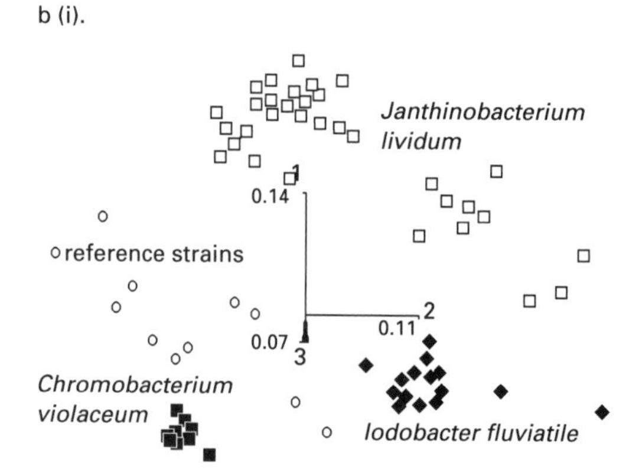

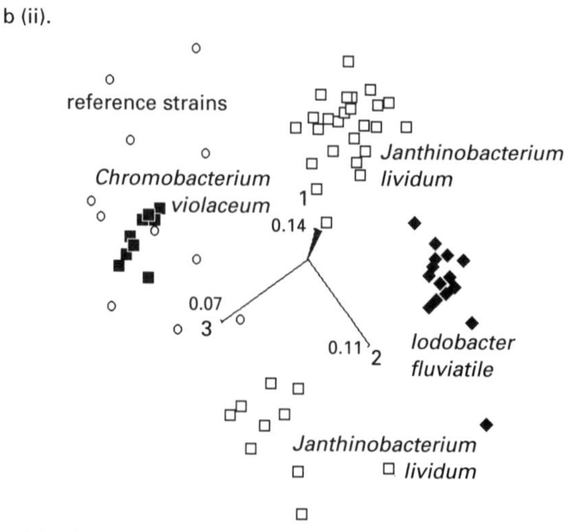

Fig. 4.5 a. A simplified phenogram of some violet-pigmented, Gram-negative bacteria. b. Rotated principal coordinate plots of the first three vectors of an analysis of some violet-pigmented, Gram-negative bacteria. Panel (ii) is a rotated view of panel (i) so that it is viewed from above.

Interpretation

The classification process becomes somewhat subjective at this stage as the taxonomist tries to apply artificial and poorly defined concepts such as species and genus to the structure revealed, on the bases of diversities within groups and spectra and distances between them, and with respect to the reference strains.

Once **phenons** (groups established by phenetic methods) have been discovered they can be defined by their characters and distinguished from others by statistical methods, as described in Chapter 5. At present, however, satisfactory and absolute criteria for establishing taxonomic rank are lacking; as noted in Chapter 1, the stringent use of genomic data for this purpose would cause problems.

The resemblance levels at which species and genera should be separated will depend upon the strain collection, character set, resemblance coefficient and clustering method used, and it may not be possible to define ranks by ruling lines across phenograms or drawing circles on ordination plots; taxonomists must be pragmatic and flexible, and recognize that their results may raise questions as well as answer them.

The problem is illustrated by the phenogram and rotated ordination plots in Fig. 4.5 which summarize the results of a larger study of violet pigmented bacteria. *Chromobacterium violaceum* strains form a compact cluster at the 85% *S* level, and all but two, atypical or outlying, strains of *C. fluviatile* have merged by 80% *S*. The groups are well separated and with the support of genomic data, a new genus, *Iodobacter*, was established for the latter species. *Janthinobacterium lividum* requires further study however; it presents as a heterogeneous species or spectrum in the phenogram with an S range of 45% (38% if outliers are omitted), yet most of the strains fall into two well separated groups in the ordination. Although the structure could be a reflection of the numbers of representatives included, the full data and other studies do not support this idea.

As a conclusion to much painstaking analytical work the interpretative approach outlined may seem a disappointment, but it should become apparent from the later chapters of this book that, lumpers and splitters notwithstanding, bacterial taxonomists tend to agree on what they mean by species and genera.

The interpretation of phylogenies and the effects of error upon them are considered in Chapter 6.

5 Identification

No matter how hard he struggles toward . . . the goal of building a system capable of identifying and grouping all kinds that exist, . . . the bacterial taxonomist is always painfully aware of the short-comings of his efforts and of the enormous amount of work remaining to be done. At best, he can only contribute to a progress report.

Ruth E. Gordon, William C. Haynes & C. Hor-Nay Pang, 1973, *The Genus Bacillus*

Introduction

The efforts of taxonomists are relevant to every kind of bacteriologist: pathologists, biotechnologists, ecologists and molecular geneticists all benefit from a better understanding of the relationships between bacteria and either need to do identification work routinely or occasionally, or rely upon the accuracy of identifications made by others.

Unless similar organisms have already been characterized, classified and named, a new isolate cannot be identified. Identification is the practical application of classification and nomenclature, which it logically follows and depends upon, and it is likewise based upon the determination of relationships between organisms. It differs, however, in its objective and approach, which are to put a name to a strain by assigning it to a previously established and named group, on the basis of a small number of characters which may be weighted.

Identification is a very exacting exercise, more so than classification, and deals in probabilities. Since organisms are never identical, identification strictly speaking, cannot be achieved; one seeks the taxon to which the unknown strain probably belongs, but like the classification upon which it is based, the answer cannot be final. Failure to identify may be owing to one or more of the following reasons (which show some overlap): (i) inaccuracy in published description or database; (ii) error in characterizing unknown strain; (iii) inadequacy of original classification; (iv) inadequate characterization of unknown strain; (v) discovery of an atypical organism; (vi) discovery of a representative of a new taxon.

The use of strictly standardized, reproducible tests is essential to minimize the first two of these problems. The last two are particularly common in environmental studies and although the range of organisms encountered in medical microbiology is much smaller, new taxa of importance are regularly described: *Legionella*, *Borrelia burgdorferi*, and *Helicobacter pylori* for example. Experience plays a very important part, of course: an isolate whose identity may have eluded the staff of a routine laboratory despite considerable effort might be recognized at a reference laboratory on the basis of a few simple observations.

A good identification system will have the following features, some of which are conflicting and make compromise necessary.

1 Reliability: the most important requirement, but one that the artificiality of taxonomy and test inconsistencies can make difficult to achieve.

2 Convenience: characters should be easy to determine, and drawbacks such as lengthy preparation and complicated interpretation may lead to disuse. Commercial systems should be easily available and have long shelf lives.

3 Rapidity: to be of value in the majority of fields, especially clinical and industrial, an early answer is required; with many modern systems, however, initial cultivation and purification of the organism are the slowest steps.

4 Relatively few tests (in comparison with the parent classification study). With large and heterogeneous groups, and those showing much variation within taxa, small numbers of characters may result in poor precision and failure to recognize atypical strains.

5 Flexibility and versatility: ability to cope with atypical strains and those with special growth or other requirements, and to accommodate new taxa.

6 Cheapness (in terms of materials and time): this is usually an important consideration.

Approaches to identification

A wide variety of techniques and test procedures is used for the identification of different bacteria because schemes for individual groups have been developed independently of one another and no one approach is universally applicable. None the less, the majority of schemes require that the organism be isolated in pure culture; those that do not may entail the sacrifice of the isolate. As a selective medium may suppress the growth of contaminants within or near discrete colonies of the chosen organism, restreaking on a non-selective medium is essential; even then some chain-forming or slime-producing strains may be difficult to free from contamination. When results are required urgently, as in a clinical laboratory, the non-selective medium is often inoculated at the same time as the identification tests as a purity check. Gross contamination may be revealed by microscopy if the chosen organism does not show much morphological variation and any contaminants have different morphologies or staining reactions to it.

Characters, chosen because they show constancy within taxa but variation between taxa, will often be a subset of those used for classification. Obtaining a pure culture is only the first standardization step in identification; the characterization tests must be strictly specified and standardized, just as they were for classification. Numerical taxonomies can tolerate some test inconsistency, but for identification work all tests should give highly consistent results (and not only in the expert hands of the parent laboratory's workers) as a single error may lead to misidentification.

Identification is a process of elimination, and the source, isolation, purification, and microscopy of the organism will furnish useful data about, for example, its general metabolism, oxygen requirements and morphology, enabling the choice of a small set of tests or an appropriate commercial kit. Failure to identify is often caused by errors in simple initial observations such as shape, motility and Gram reaction, which result in inappropriate tests being subsequently applied. Even when using chemotaxonomic systems with broad databases, it is prudent to take account of the whole range of available information. Only when identification fails repeatedly for several similar strains, which have been characterized by tests that have been proved reliable with positive and negative control cultures, should the presence of a new taxon be considered and a classification study planned.

Characterization

The categories of characters used for identification are the same as those for classification (Table 2.1), but some different examples have been given in Table 5.1 to emphasize two points. First, in most cases the initial observations use methods that have seen little change in half a century and may represent all but three of the categories shown in the table; for example, a Gram-negative, motile rod isolated on deoxycholate citrate agar and presumptively identified as *Salmonella* will have been subjected to cultural

Table 5.1 Categories of characters used for identification

Category	Examples
Cultural	Colonial morphology and reactions on differential media
Morphological	Cell shape, Gram reaction, motility
Physiological	Oxidation/fermentation test, growth at 37°C
Biochemical	Acid from carbohydrates, decarboxylase tests; miniaturized rapid test kits with automatic reading and interpretation
Nutritional	Sole carbon and energy sources; miniaturized rapid test kits with automatic reading and interpretation
Chemotaxonomic	Electrophoresis of radiolabelled proteins with automatic scanning and interpretation, automatic analysis and interpretation of whole-organism fatty acids
Serological	Coagglutination, immunofluorescence, enzyme-linked immunosorbent assays
Inhibitory tests	Growth on selective media, inhibition by antibiotic disks
Genotypic	Nucleic acid probes

(characteristic colonies), morphological (microscopy), physiological (growth at 37°C in air), biochemical (failure to ferment lactose), nutritional (utilization of citrate) and inhibitory tests (growth in presence of bile salt). Second, many biochemical and nutritional tests and several serological and genomic approaches are available in rapid miniaturized kits, some of which feature automatic interpretation, and some chemotaxonomic techniques have been successfully developed for automatic processing, reading and interpretation.

Traditional methods

Biochemical, nutritional and physiological characterization tests, usually carried out in bottles and tubes of solid or liquid media and on plates, have been developed and modified since the earliest days of bacteriology, largely because of the limited value of cultural and microscopical features in identification. All five such approaches are still widely used for identification (and classification, as outlined in Chapter 2) and they are variously described as traditional, classical or conventional methods (although several modern, and by implication unconventional, methods have been widely used for over 20 years!).

Traditional tests may sometimes be inconvenient to prepare, difficult to standardize and interpret, poorly reproducible, and slow to perform, but they have the advantage of cheapness. They may be also the only methods available for poorly classified groups and for those of little or no clinical or industrial importance, so that the development of modern rapid methods has not been commercially worthwhile. It must be emphasized that even with important groups, for which modern methods are available, traditional tests continue to be valuable in the early stages of identification and may, especially in the hands of experienced workers, allow a very rapid and accurate result with little further effort. This is particularly true in clinical laboratories where cultural and microscopic characters often suggest identities that can be confirmed with single tests (examples are the coagulase test for *Staphylococcus aureus* and optochin sensitivity for *Streptococcus pneumoniae*).

Composite media

For some groups of organisms which are not so easily distinguished the inconvenience of having to prepare and inoculate several different test media (Fig. 5.1) has been reduced by building several tests into plate media for isolation or into tubed composite media for the testing of pure cultures. Numerous selective and differential media have been devised, especially for clinical and food bacteriology (an example is polymyxin egg-yolk mannitol bromthymol blue agar, on which colonies of *Bacillus cereus* may be distinguished from those of other *Bacillus* species by their egg-yolk reaction and failure to ferment mannitol). Many types of tubed composite media have been devised but their tests may interfere with each other and few remain in wide use; they include Kligler iron agar, triple sugar iron agar

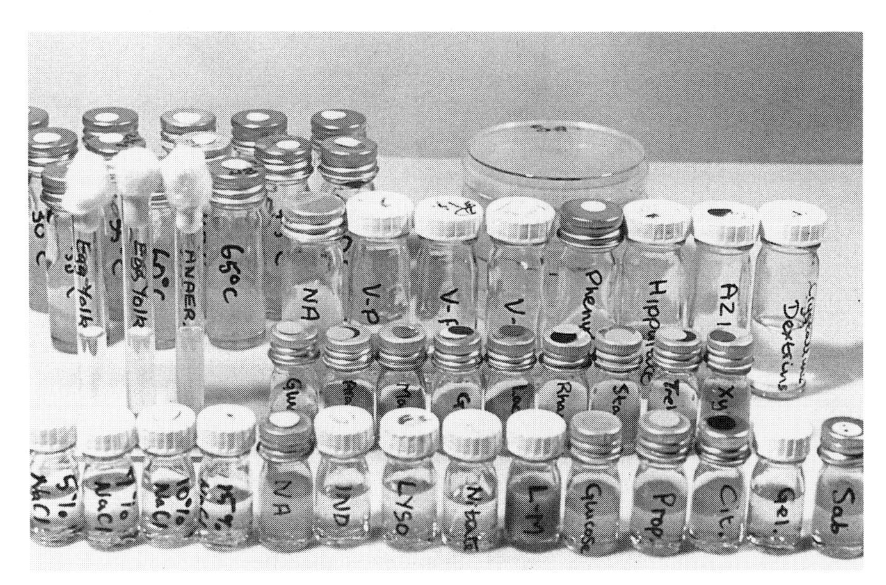

Fig. 5.1 Media used for characterizing a strain of *Bacillus* by conventional methods. Tests include: maximum and minimum growth temperatures, starch and casein hydrolysis, blackening of tyrosine, dihydroxyacetone production, tyrosine decomposition, egg-yolk reaction, anaerobic growth, catalase reaction, Voges–Proskauer test and final pH in V–P broth, phenylalanine deamination, hippurate hydrolysis, growth in glucose azide broth, crystalline dextrin formation, acid from carbohydrates, growth in NaCl broths, microscopic appearance, indole production, lysozyme resistance, nitrate reduction, litmus milk, blackening of glucose, utilization of citrate and propionate, gelatin liquefaction and growth at pH 5.7.

and Kohn's two-tube medium (Fig. 5.2), all of which are used to identify members of the family *Enterobacteriaceae* on the basis of multiple carbohydrate fermentations, hydrogen sulphide production and, in the last case, motility and indole production as well.

Micromethods

As growth of the organism is required, tests in composite media still suffer from slowness. This problem has been tackled by adding heavy inocula to small volumes of individual media or to enzyme substrates. The former approach, like traditional methods, can reveal preformed and induced enzymes as some rapid multiplication briefly occurs, whereas the latter shows only preformed enzymes. The enzyme substrates may be chromogenic or fluorogenic so that a separate indicator system is unnecessary. The reactions do not always correspond well with the standard methods so that different databases must be used, but results may be obtained within minutes or hours rather than hours or days. The rapidity of these tests may be devalued by the need for inocula heavier than primary isolation plates can provide, but with detection of enzyme activity by sensitive instrumental methods such as fluorometry one colony may be sufficient to inoculate a battery of tests; the rapidly developing technology of biosensors will undoubtedly make a major contribution in this area. The wide commercial

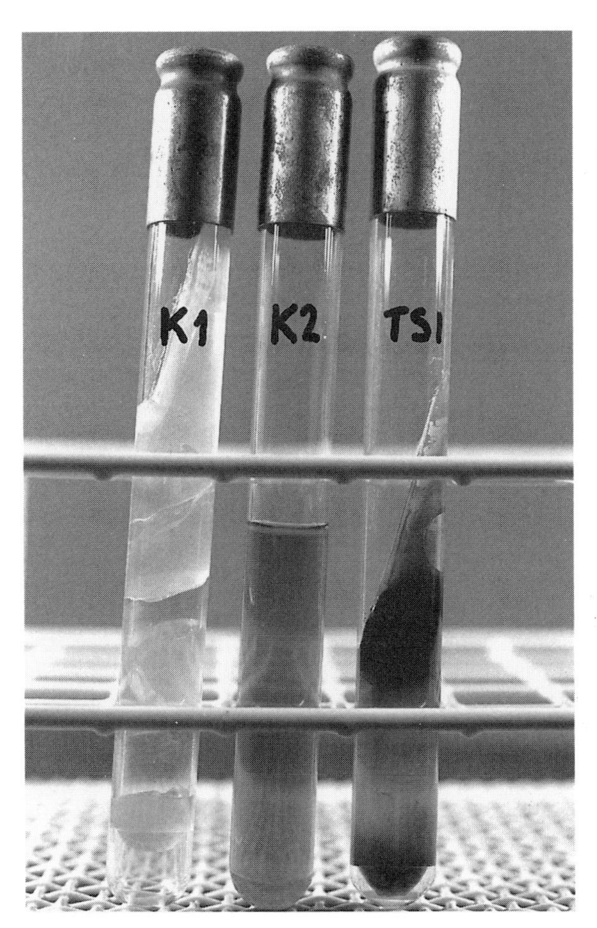

Fig. 5.2 Composite test media for enterobacteria. From left to right. Kohn's Medium No. 1 (K1) shows fermentation of glucose and mannitol (medium turned from red to yellow) and gas production (bubbles disrupting the agar); urease production would be indicated by a red reaction throughout. Kohn's Medium No. 2 (K2) shows motility (whole medium turbid), fermentation of sucrose and/or salicin would be indicated by medium turning from blue-green to yellow; indole production can be detected by suspending a test paper from the rim of the tube. Triple sugar iron agar (TSI) shows H_2S production (black deposit) and fermentation of glucose (yellow butt with gas) but not of lactose or sucrose (slope remains red). Results are typical for a *Salmonella* species.

availability of miniaturized test kits may have inhibited the widespread adoption and further development of 'home-grown' micromethods, but for many environmental studies (such as that of marine bacteria) the latter can be superior.

Commercial identification kits

Commercial miniaturized kits have long shelf-lives and offer strictly standardized, rapid tests in uniform arrangements, which are simple and quick to inoculate and give consistent results. Many clinical laboratories use such systems routinely and believe the convenience to be worth the expense which may, in some cases, be compensated for by shorter hospital stays and reductions in labour costs.

The earliest systems were developed for the identification of enterobacteria, but kits for other groups, including anaerobes, non-fermenting Gram-negative bacteria, *Bacillus*, *Lactobacillus*, *Neisseria*, *Staphylococcus* and *Streptococcus*, soon followed. Various formats are used, including dehydrated media and substrates in plastic strips of microtubes or microtitre plates, or paper disks for addition to welled plates, and with inoculation by pipette; or agar media, either in conventional plates for multipoint inoculation and simultaneous testing of many organisms, or in multicompartmented tubes inoculated by integral needle; some examples are shown in Fig. 5.3.

Results are commonly converted to profile numbers by the **octal coding system** as shown for an API 20*Enterobacteriaceae* strip in Fig. 5.4: positive reactions in the first, second and third test in each triplet score 1, 2 and 4, respectively. The seven-digit number so generated leads to an identification, or recommendation for further tests, in the manufacturer's manual or from an extensive and continuously expanding computer database, which is accessible by telephone. As these systems usually require overnight incubation at least, some manufacturers produce faster kits, which contain selections of tests developed from one or both of the two kinds of micromethods described above, and which give results in 4 hours.

Another factor which may contribute to test inconsistency is subjectivity in interpretation, but advances in instrumentation and computing have made possible the automatic reading of test kits, thus removing this source of error. The API ATB system (Fig. 5.5) is one of several systems now available; it uses special 32-well strips containing dehydrated media and enzyme substrates which may, with suspensions of standard opacity, be inoculated automatically. After incubation (which is only 4 hours for enterobacteria) and addition of reagents, the reactions are read by spectrophotometry and compared with the database, and the identifications printed. The Biolog system uses 95 carbon utilization tests (see Chapter 2) in Gram-negative and Gram-positive panels, and results may be interpreted using a microplate reader; such large numbers of tests give great versatility so that the main constraint of the system is the size of its database.

Chemotaxonomic methods

As later chapters show, chemotaxonomic methods are of particular value for the identification of organisms that are unreactive in, or inseparable by, conventional tests. Descriptions of many techniques have emerged from the laboratories of taxonomists and several systems have been developed commercially.

Gas–liquid chromatography has been widely used to determine the volatile and non-volatile acid fermentation products of anaerobic bacteria. Rapid, accurate and reproducible results can be obtained with inexpensive instruments to give definitive identifications of many clinically important

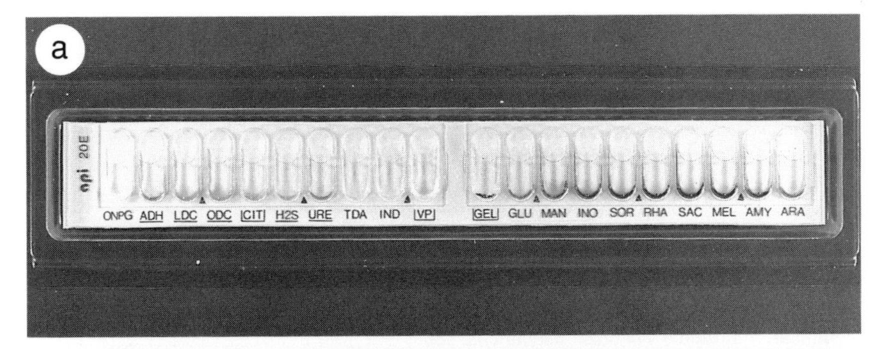

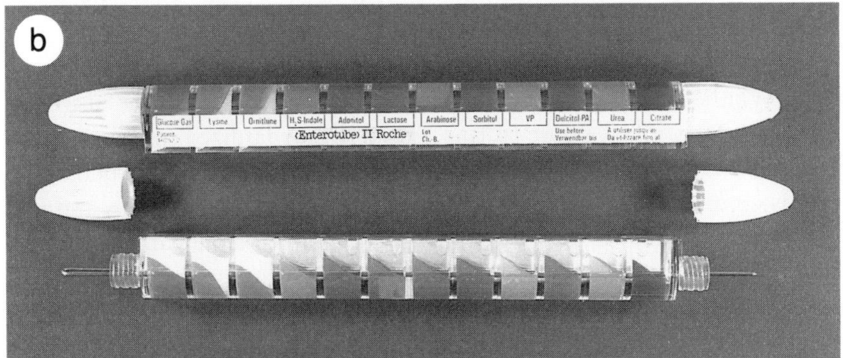

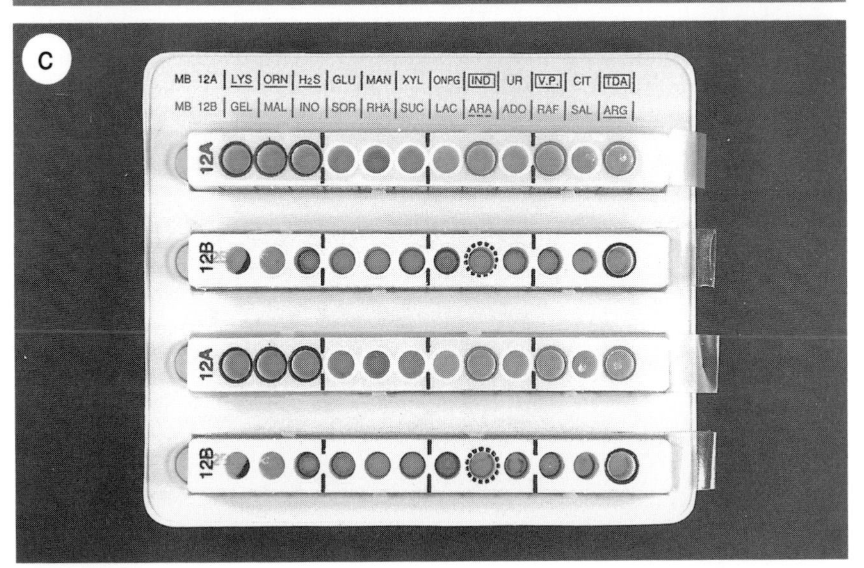

Fig. 5.3 Identification kits for enterobacteria. a. The API 20E kit comprises a strip of 20 tubes containing dehydrated test media which are reconstituted by adding a suspension of the test organism in water; the moistened incubation chamber is covered with a lid to prevent the tubes drying out. b. The Roche Enterotube II contains 12 sloped media that are inoculated by picking off a colony with the integral needle and withdrawing the needle through the compartments; the needle is partially replaced to maintain anaerobic conditions, the remainder being snapped off, while the other 8 compartments are punctured to render them aerobic. c. The DP Microbact 12A and 12B systems comprise strips of 12 wells containing dehydrated test media that are reconstituted by adding a suspension of the test organism in saline; the wells are then resealed with their cover strips. In all three kits reagents are added as appropriate after incubation.

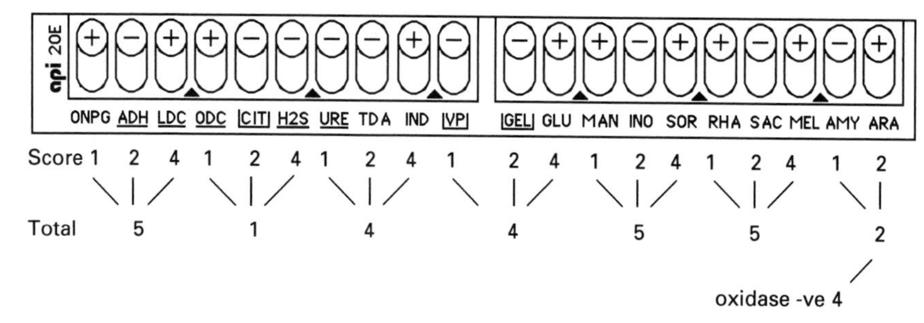

Identification profile 5 144 552 = *Escherichia coli*

Fig. 5.4 The octal coding of an API 20E strip.

taxa. Whole-organism fatty acid analysis by GLC is another powerful method which, given a pure culture grown in standardized conditions, can yield an identification from a few colonies in less than 2 hours, such a system is commercially available. Other lipids may also be of value, and isoprenoid quinones, for example, have proved to be particularly useful for separating *Legionella* species.

The AMBIS system is based upon protein pattern matching: proteins are labelled with [^{35}S]methionine and separated by automated electrophoresis using standardized precast gels. These are then scanned two-dimensionally for beta emission and the patterns compared with a database. With some groups PAGE methods can be valuable for typing down to a level equivalent to serovar.

Pyrolysis mass spectrometry is potentially of great value in bacterial identification, and in typing for epidemiological studies, as samples can be

Fig. 5.5 The API ATB System for automated bacterial identification and antibiotic sensitivity testing. After incubation, the test strip is placed in the reader on the left; this records the colour reactions and transmits the information to the computer in the centre. Hard copy such as a hospital patient report is generated by the printer on the right.

analysed in a few minutes. It has found several routine applications, including the study of mycobacteria and the screening of actinomycetes. Pyrolysis of single cells by laser microprobe offers even shorter identification times as organisms do not have to be cultivated.

Two other rapid methods undergoing development are Fourier-transform infrared spectrometry and ultraviolet resonance Raman spectrometry; their principles are outlined in Chapter 2.

Serological tests

Serological cross-reactions between unrelated organisms occur frequently. Serological tests are therefore chiefly used to confirm presumptive identifications and for typing purposes, and are widely used in clinical laboratories where a particular organism might be sought or epidemiological tracing is required. They are increasing in popularity because of their rapidity, ease of use, specificity and sensitivity, and kits for a wide range of organisms are available. They are usually based on labelled polyclonal and monoclonal antibody methods such as latex particle agglutination, coagglutination, immunofluorescence and enzyme-linked immunosorbent assay (ELISA).

Inhibitory tests

As indicated above, selective media can be very helpful in narrowing down the range of taxa to be considered. Several widely used screening tests use disks impregnated with antibiotics or other inhibitory agents; the sensitivity of *Streptococcus pyogenes* to bacitracin and of vibrios to 0/129 are familiar examples. One semi-automated identification system identifies enterobacteria and non-fermentative Gram-negative rods entirely on the basis of differential growth inhibition by antibiotics, chemotherapeutics, dyes and other chemicals such as sodium azide and thallous acetate; profiles are interpreted by discriminant analysis (see Distance Models below).

Genotypic methods

Nucleic acid hybridization is useful with some bacterial groups, especially where the spirit of classification has been splitting rather than lumping, but even with rapid methods it is too complicated and time consuming for most purposes. More appropriate for routine use, and of particular value for organisms that do not have diagnostic antigens or are not readily cultivated, are **nucleic acid probes**. These are short lengths of labelled ssDNA, specific for a single gene or locus, or pieces of RNA, that hybridize with complementary strands of nucleic acid in the specimen, so allowing the detection of, for example, a species or pathogens within a species. Such probes are now widely used in clinical and industrial laboratories and several kits are commercially available.

Probes may be labelled in various ways. Biotin, which covalently binds to DNA without affecting its hybridization properties, is used in several detection methods. Avidin, a protein that binds tightly to biotin and to which various markers such as enzymes, fluorescent dyes and ferritin may be attached, is added after hybridization. Greater sensitivity is obtained if the probe is labelled with a radioisotope such as ^{32}P, but because the photographic detection methods commonly used are inconvenient, highly sensitive non-isotopic systems are being developed. Using an immunoassay approach, antigenic probes may be detected by enzyme activity or by time-resolved fluorometry and the latter method can also be used to detect probes labelled directly with rare earth chelates.

The sensitivity of probing can be greatly enhanced by preceding it with PCR amplification (potentially by up to 10^{12}-fold) of the target region. This method is particularly attractive to clinical laboratories, as small numbers of serious pathogens such as *Mycobacterium* species, whose isolation may be slow and difficult (and sometimes impossible), can be detected within hours rather than days or weeks. An alternative way of increasing sensitivity is to probe for 16S rRNA, which is abundant in cells; probes may be fluorescently labelled so that single cells can be detected by microscopy.

Another sensitive and accurate genotypic approach, which does not require probes to be developed, is restriction endonuclease analysis of chromosomal DNA. This is proving useful for identification within some of the more intractable groups such as *Leptospira*.

Interpretation

As seen earlier, if the bacteriologist has a shrewd idea of an isolate's identity a small set of tests, or perhaps a specialized kit, may be used for rapid confirmation. However, when embarking upon the exercise without preconceived ideas, or when a hunch has proved wrong, identification is approached in a sequential or stepwise fashion and the results of the initial tests determine what set of tests is used next. This is more economical in time and materials than what Cowan & Steel called the 'blunderbuss method' in which numerous tests are performed simultaneously in an unstructured way and the results compared with those in reference texts — with this approach, crucial tests may have been omitted or performed by unsuitable methods and many of the tests might be redundant.

Diagnostic keys

Diagnostic keys have been used by biologists for many years, but they can be difficult to construct without the aid of a computer. The commonest type is the dichotomous key in which each step is based upon a test with two possible results (Table 5.2), the final step being to confirm the identification by comparison with a detailed description of the taxon, usually in tabular form. The tests used must be reliable, give unequivocal positive and negative results, and be consistent for each taxon otherwise a single test error would lead to misidentification or failure to identify. As the presence or absence of a single property is sufficient to exclude the unknown from a

Table 5.2 A dichotomous key for violet-pigmented bacteria

1	Growth at 37°C	*Chromobacterium violaceum*
	No growth at 37°C	2
2	Aesculin hydrolysis	*Janthinobacterium lividum*
	Aesculin not hydrolysed	3
3	Anaerobic growth	*Iodobacter fluviatile*
	No anaerobic growth	Atypical *J. lividum*

taxon, such keys are monothetic and the characters are weighted. Where a character is variable for a particular taxon, however, both positive and negative results may be allowed so that the identification is reached by two routes and the test is, in effect, skipped. Unfortunately, characters sufficiently constant for inclusion in keys are often widely shared and only of value for separating the higher taxonomic ranks. In recent years relatively few keys have been devised for bacterial identification to species level; their sequential nature makes them slow, and between-strain variation and test inconsistency make them too unreliable in comparison with other methods.

Multiple keys or **polyclaves**, in which tests are considered in any order and taxa eliminated in turn, require similar kinds of tests to dichotomous keys but are more flexible in use. They were formerly prepared as punched card systems but are now stored in a computer.

Diagnostic tables

A table has several advantages over a key: it can be used sequentially or several characters can be considered simultaneously, between-strain variation can be shown, the data may be transferable to computer for numerical identification, it contains more information on each taxon so that it is better suited for publication, and because patterns of characters are used it is a polythetic method, and the occasional deviant result is less likely to mislead.

Positive and negative reactions for 95% or more of strains are usually indicated by + and – respectively, v or d (for variable or differing results) are often used to indicate between-strain variation, and various other symbols may be used for further divisions (Table 5.3a) and weak or delayed reactions. A table showing percentages of positive results (Table 5.3b) is preferable as it contains more information, and if numbers of strains are shown the user can judge the value of the data and modify them as further isolates are characterized. Sometimes +/– and percentage data are shown in adjacent columns. Regardless of how the data are presented, tables can be difficult to read objectively if many variable results are included for individual (perhaps poorly defined) taxa. This is illustrated by the first three columns of Table 5.3a and b: column AB contains few diagnostically useful responses and the character correlations apparent in the other two columns, A and B, are concealed.

Table 5.3 Diagnostic tables for violet-pigmented bacteria

a.	Group*				
Property	AB	A	B	C	D
1 Growth at 37°C	–	–	–	–	+
2 Anaerobic growth	–	(–)	–	+	+
3 Aesculin hydrolysisv	+	–	–	–	
4 Acid from trehalosev	–	+	+	+	
5 Lactate utilization	+	+	(+)	–	+

b.	% Positive in group* (number of strains)				
Property	AB (82)	A (68)	B (14)	C (53)	D (9)
1 Growth at 37°C	0	0	0	0	100
2 Anaerobic growth	4	5	0	100	100
3 Aesculin hydrolysis83	100	0	0	0	
4 Acid from trehalos	18	1	100	100	100
5 Lactate utilization	100	100	93	0	100

*Groups: AB, *Janthinobacterium lividum*; A, typical *J. lividum*; B, atypical *J. lividum*; C, *Iodobacter fluviatile*; D, *Chromobacterium violaceum*.
+, 95–100% strains are positive; (+), 85–94% strains are positive; v, 15–84% strains are positive; (–), 5–14% strains are positive; –, 0–4% strains are positive.

To use a table, the results of the unknown are written in order on a strip of paper and compared with each of the taxa in turn until the best fit is found; if many tests and taxa are to be considered the table can be broken down into several smaller tables, so that the identification becomes a two- or three-stage exercise and is partly sequential and partly simultaneous. This approach has been used successfully in several practical reference books.

Numerical identification

Computer-based simultaneous methods offer several advantages over keys and diagnostic tables. Numerical approaches allow measures of certainty to be attached to identifications, yet they will often work with fewer tests and when the unknown gives some deviant results. Also, databases can contain numerous characters and be updated as required — continually if they are on-line. Developments in artificial intelligence allow identification with incomplete databases compiled from the literature by examining unknowns from different points of view. Numerical identification systems may be evaluated by comparing DNAs from identified organisms with reference DNAs from the type strains of the appropriate species.

The two kinds of numerical identification methods most widely used in bacteriology, conditional probability and distance models, are easily constructed from tables giving percentages of positive results for binary tests. Percentages are divided by 100 to give proportions of positive results, and values of 1 and 0 are set to 0.99 and 0.01, respectively, to allow for atypical results and prevent multiplication by zero (Table 5.4).

Conditional Probability Models. If proportions are treated as probabilities, **conditional probability** or **likelihood** models may be applied to diagnostic tables such as Table 5.4 to give **probabilistic polyclaves**. If a taxon shows a probability of 0.99 for a test, this is taken to mean that the chances of belonging to it are 0.99 for an unknown with a positive result and 0.01 for one with a negative result; in the latter case the probability has been obtained by subtracting the value in the table from unity. For several characters the likelihood of a taxon giving the same results as the unknown is calculated by multiplying the probabilities together, and as the number of characters increases the likelihood of a misidentification decreases faster than that for a correct identification. If each likelihood is then divided by the sum of likelihoods for all the taxa being considered, a **normalized identification score** is obtained, and this approaches unity where the unknown identifies closely with one taxon but little with any of the others. The computer can be programmed to indicate the presumptive identification and next best candidates and scores and, if a satisfactory identification cannot be achieved, which further tests should be performed. This very powerful method was developed by Lapage and colleagues at the NCTC in London, and most commercial kits use a combination of this approach with direct pattern matching methods for profile numbers.

For the data in Table 5.4, the unknown is compared with the four taxa A, B, C and D as follows:

$$\begin{aligned}
&\text{A}\ (1-0.01)\times(1-0.05)\times0.99\times(1-0.01)\times0.99 = 0.91256621\\
&\text{B}\ (1-0.01)\times(1-0.01)\times0.01\times(1-0.99)\times0.93 = 0.00009115\\
&\text{C}\ (1-0.01)\times(1-0.99)\times0.01\times(1-0.99)\times0.01 = 0.00000001\\
&\text{D}\ (1-0.99)\times(1-0.99)\times0.01\times(1-0.99)\times0.99 = 0.00000001
\end{aligned}$$

$$\text{the sum of the likelihoods, } \Sigma = 0.91265738$$

Table 5.4 Probability table for numerical identification

	Group*				
Test	A	B	C	D	Unknown
1	0.01	0.01	0.01	0.99	–
2	0.05	0.01	0.99	0.99	–
3	0.99	0.01	0.01	0.01	+
4	0.01	0.99	0.99	0.99	–
5	0.99	0.93	0.01	0.99	+

*Groups and tests as in Table 5.3.

so that the normalized identification scores are:

A 0.91256621/Σ = 0.99990010
B 0.00009115/Σ = 0.00009987
C 0.00000001/Σ = 0.00000001
D 0.00000001/Σ = 0.00000001

and, with a score exceeding 0.999, so that the scores for all the other taxa are below 0.001, the unknown is identified as a typical strain of taxon A (*J. lividum*).

Distance Models. In the **taxon-radius model** each taxon is regarded as a cluster of strains in hyperspace and character scores are treated as distances in that space. If it is assumed that the taxa are roughly **hyperspherical** (sphere-shaped in hyperspace) and the scatter of strains in each is similar to a normal distribution, an envelope of **critical radius**, *r*, containing the majority of strains and representing a confidence limit can be drawn for each one about its centre of gravity (**centroid**); the radii of homogeneous taxa will be small and those of heterogeneous taxa large. Each taxon is therefore represented as a hypersphere of given size and position in the multidimensional identification space, and the characters of an unknown will determine its position in that space. A typical strain will lie within an envelope and be identified, an atypical strain will lie just outside an envelope, an intermediate strain will lie midway between two envelopes, and a strain representing a taxon not included in the scheme will lie at some distance from all the envelopes and remain unidentified (Fig. 5.6).

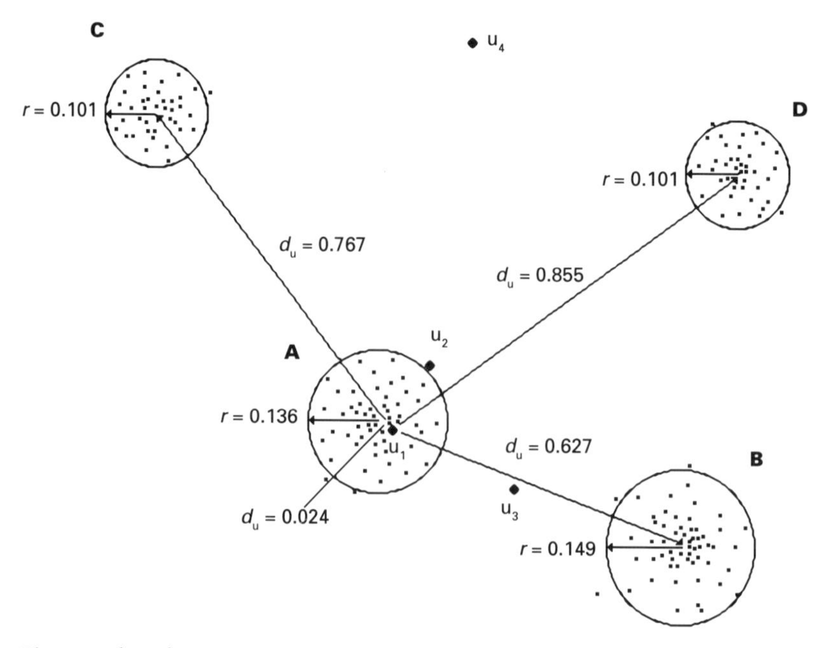

Fig. 5.6 Identification using a taxon-radius model; u_1, unknown identified as member of A; u_2, an atypical strain of A; u_3, an intermediate between A and B; u_4, unidentified.

The procedure may be illustrated using the data in Table 5.4. The mean distance d of the members of a taxon from their centroid is approximated using the formula

$$d = \sqrt{[1/n \, \Sigma(P(1 - P))]}$$

where n is the number of characters and P is the probability of a positive result. The approximate standard deviation of d, s, is given by

$$s = \sqrt{(d^2/2n)}$$

The critical radius is chosen to include a theoretical proportion of the strains such as 99% and is given by

$$r = d + (ks)$$

where k is obtained from a table of one-tailed normal distribution functions; for a probability of 0.99, as suggested, it is 2.33. The distances d_u of the unknown from the taxon centroids are then calculated from

$$d_u = \sqrt{[1/n \, \Sigma(X - P)^2]}$$

where X is the unknown's test result; for example, with group A the values of $X - P$ in tests 1 to 5 are −0.01, −0.05, 0.01, −0.01 and 0.01 respectively. The distance values for the example are shown in Table 5.5 and it can be seen that the unknown lies well within the envelope of A. This example is based upon binary characters but quantitative character means, scaled to a 0,1 range to avoid weighting, can also be used.

If much character correlation exists, the taxon-radius model will fail because the taxa are elongated so that critical radii cannot be measured. It will also fail if intermediate strains are common so that the taxa overlap. Conclusive identification may not then be possible and the problem becomes one of discrimination. Methods of **discriminant analysis** distinguish between two preclassified populations (i.e. taxa) by finding a boundary of equal probability between them; an unknown is assigned to the taxon on the same side of the boundary as itself (Fig. 5.7).

Canonical variates analysis enables discrimination between more than two groups by searching for gaps between them in an ordination. Relatively few taxa can be accommodated and ideally their sizes, shapes and

Table 5.5 Distance values for taxon-radius model

	Group*			
Value	A	B	C	D
Mean distance (d)	0.132	0.144	0.099	0.099
Standard deviation (s)	0.002	0.002	0.001	0.001
Critical radius (r)	0.136	0.149	0.101	0.101
Distance of unknown to centroid (d_u)	0.024	0.627	0.767	0.855

*Groups as in Table 5.3

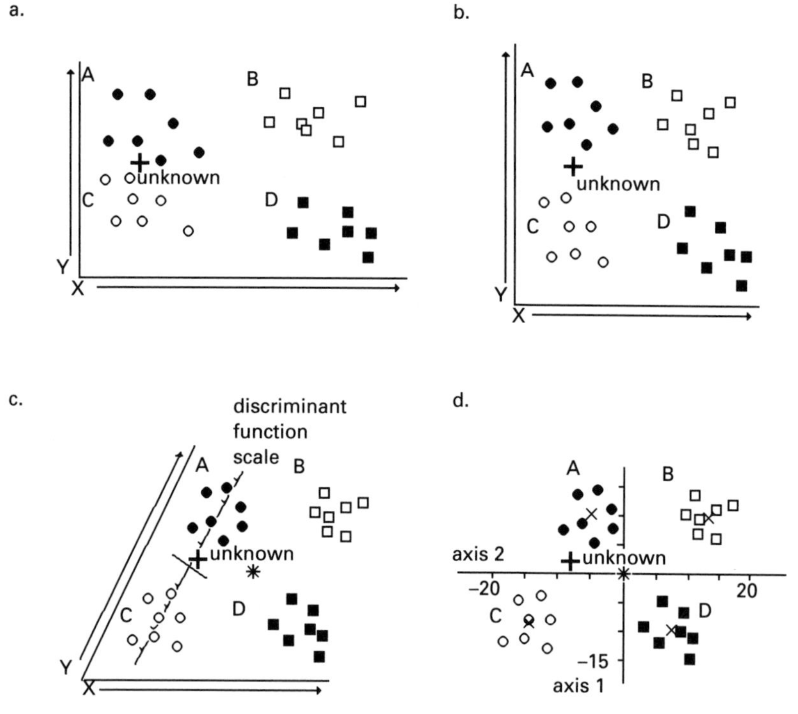

Fig. 5.7 An illustration of the steps in discriminant analysis for four groups of organisms A, B, C and D, and two quantitative characters. The identity of the unknown is unclear from the untransformed data (a), so the character axes X and Y are contracted and expanded respectively (b), and the angle between them altered (c) so as to make the groups as spherical as possible. As the unknown lies above the central bar on the discriminant function scale it can be identified as a member of group A. By rotating the system around its centre of gravity (shown by the asterisk), the main axes of variation can be found (d) and the unknown is found to lie closer to the centre (marked X) of group A than to the centre of group C.

orientations should be similar. The mathematics is complicated, but the method essentially maximizes the dispersion between the groups by squaring Euclidean distances, and alters the angles between the character axes so as to take character correlations into account by weighting and make as many as possible of the groups hyperspherical. Distances between an unknown and the taxon centroids can then be measured in the transformed space. Discriminant methods are most appropriate for quantitative characters and have occasionally been used for analysing chemotaxonomic data such as those from pyrolysis mass spectrometry.

6 Evolution and the archaea

The study of bacterial evolutionary relationships is central to the historical account of life on this planet.

Carl R. Woese (1987) *Bacterial Evolution*

Introduction

Haeckel's phylogenetic tree of 1866 had three main branches to represent animals, plants and microorganisms evolving from a common ancestor, but *Vibrio* was the only recognizable bacterial group included. Information gathered during the subsequent 100 years led Whittaker to recognize five kingdoms which contained the animals, plants, algae, bacteria, fungi and protozoa, but by this time the existence of two primary kingdoms, the prokaryotes and eukaryotes, was widely accepted. In the absence of fossil or genetic evidence, bacterial phylogenies were constructed on the bases of physiology and morphology, with autotrophs and cocci respectively being regarded as the most primitive forms. Contemporary theories of the origin of life, however, visualized the oceans as primordial soups and led to the inferences that the first organisms were anaerobic, fermentative, heterotrophic bacteria like clostridia or streptococci and that aerobes, autotrophs, phototrophs and eukaryotes were later developments.

More recently, shallow marine cherts from the Archaean period, about 3.5 billion years ago, have yielded well-preserved fossil microorganisms, and it has been suggested, on the basis of carbon isotope ratios of earlier sediments, that oxygenic photosynthetic bacteria existed 3.8 billion years ago. Although a surprising amount of information may be derived from microfossils, it cannot tell us much about prokaryote phylogeny. However, in the 1970s the increasing application of molecular methods to taxonomy revealed that the amino acid sequences of proteins and the nucleotide sequences of DNA and RNA contain a record of bacterial evolution. From sequencing studies of 16S rRNA in particular, by Carl Woese and colleagues at the University of Illinois and elsewhere, a comprehensive phylogeny of life on earth is emerging. The prokaryotic origins of chloroplasts and mitochondria have been confirmed, yet the eukaryotic line of descent appears to be as ancient as that of the bacteria.

Most exciting was the discovery, in the late 1970s, that there are two quite separate groups of prokaryotes; these were named **eubacteria** and **archaebacteria**, and subsequent work has shown that, in evolutionary terms, these two primary kingdoms are as distinct from one another as each is from the eukaryotes. This three-way division is not compatible with the old five kingdom arrangement of Whittaker or the prokaryote–eukaryote division and it has therefore been proposed that three **domains** (representing a new and higher taxonomic rank) the *Archaea*, the *Bacteria*, and the *Eucarya* should be recognized, corresponding with the archaebacteria, eubacteria and eukaryotes respectively (Fig. 6.1). This allows the

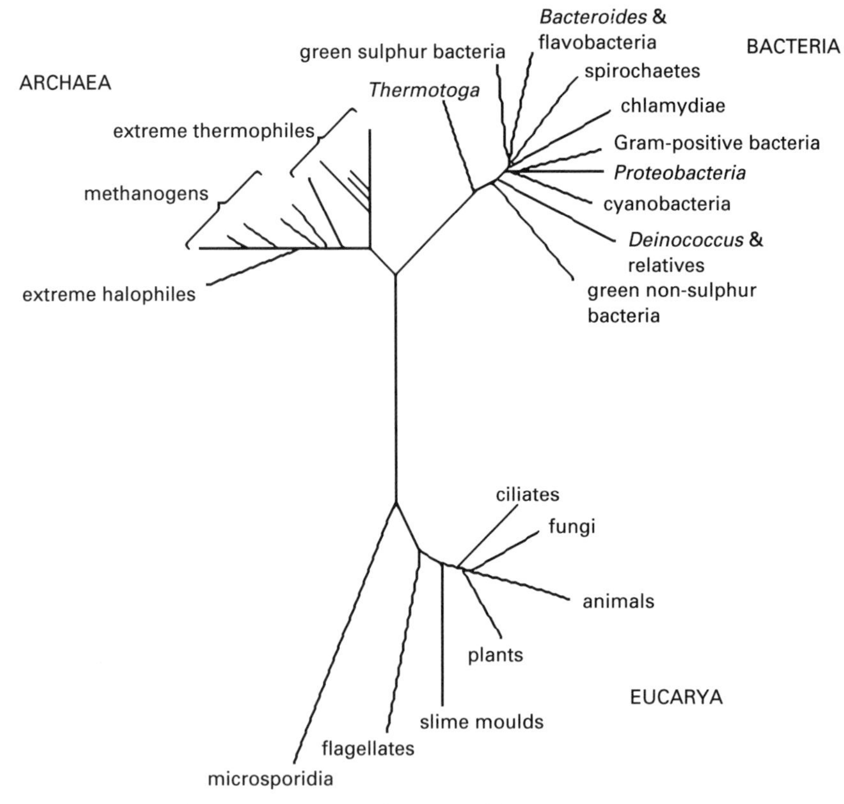

Fig. 6.1 A general phylogenetic tree determined from 16S rRNA sequence comparisons.

continued use of the kingdoms *Animalia*, *Plantae* and *Fungi*, while *Protista* is split into several kingdoms, and the bacterial divisions or phyla can likewise be given kingdom rank.

Molecular clocks

Genotypic change occurs constantly and those mutations not influenced by selective pressures become fixed in time. If all extant life on earth has descended from one ancestor by diverging paths, nucleic acids and proteins will contain sequences inherited from that ancestor and may be used as molecular clocks. Furthermore, if only limited horizontal genetic transfer has occurred and evolutionary rates have been fairly constant, organisms whose equivalent molecules show great similarity are likely to have diverged from their shared ancestor more recently than those whose molecules have fewer sequences in common.

If horizontal gene transfer between bacteria was extensive convergent evolution might be expected, but although transfers of antibiotic resistance, pathogenicity and other phenotypic characters are of undoubted importance to bacterial evolution, the few comparative studies made suggest that they are not sufficient to obscure the direct evolutionary story.

Phylogenies inferred from different data such as amino acid sequences of proteins and nucleotide sequences of 16S rRNA are often congruent.

Evolution does not proceed at a constant rate, but varies between and within lineages at different stages in their histories. Low homology between two organisms might therefore be the result of early divergence or very rapid evolution. These may be distinguished by comparing a group's sequences with those of a distantly related reference organism; if the members of the group are **isochronic** (evolving at the same rate), their homologies with the reference should be similar, and faster evolving members would be indicated by lower homologies.

Molecular clocks also run at different rates. Not only do classes of proteins evolve at widely differing speeds, rates of change within a class may also vary. Comparison of *c*-type cytochromes and of other proteins has been valuable in phylogenetic studies of eukaryotes, but is less useful for studying the evolution of prokaryotes owing to their wide biochemical diversity, and because limits determined by functional constraints on the molecules may be reached, beyond which convergent and parallel mutations become as frequent as divergent ones. The enormous scope of prokaryote evolution means that DNA–DNA hybridization methods are only useful for comparisons within genera as the genome, overall, is too fast a clock. Ribosomal RNAs are slower clocks and DNA–rRNA hybridization studies, which are quick and cheap to perform, are valuable for filling in the outer branches of phylogenetic trees, but they are not sufficiently sensitive to detect deeper branchings and their orders within kingdoms. The RNAs and other molecules must therefore be sequenced to yield their phylogenetic information but, until the late 1980s, complete sequencing was only feasible for 5S rRNA and, because of their smaller information content and other constraints, these data are of limited value for prokaryote phylogeny in comparison with 16S rRNA catalogues or sequences.

As indicated in Chapter 3, 16S rRNA is, at present, the most useful of the molecular clocks because of its universal distribution, functional constancy, large size, and the varying rates of change that occur in different parts of the molecule. These rates of change not only encompass the spectrum of evolutionary distances that are found amongst the prokaryotes, but can also allow distinction between rapid evolution and early branching. In addition to having high information contents the large rRNAs comprise many functional units. These appear, in an evolutionary sense, to be relatively independent of each other so that a major change in one hardly affects the others, yet adjacent positions within units tend to change at similar rates, and this linkage helps to conserve **signature** regions that may be characteristic of a group and is useful in determining branching orders. Some positions in the molecule appear to be invariant, others only show change between domains, and some may vary within genera. The rapidly moving hands of the clock may be moving more than two orders of magnitude faster than the slower ones which can, however, speed up disproportionately during periods of rapid evolution.

Interpretation of data

Current views of prokaryote phylogeny are dominated by 16S rRNA sequence data, but it is unlikely that this one approach, based on a small part of the genome, will answer all our questions on bacterial evolution. The extents to which convergent and parallel evolution and lateral gene transfer occur are not known, and we do not know how far molecular change is limited by structure and function relationships. Furthermore, sequence alignment is a subjective exercise; it may be difficult to decide which sites are really homologous and ambiguous alignments can occur. Problems that may occur with sequence data include non-independence of sites, lineage-dependent inequalities in rates of change, and inequalities in base substitution frequencies.

Statistical sampling error, which is greater for shorter sequences, is a major contributor to uncertainty in the branching orders of evolutionary trees, yet in many such trees the lineages branch from quite small areas. Here, as elsewhere in prokaryote classification, polyphasic taxonomies are important in maintaining flexibility and achieving a balanced picture. Indeed, anomalies revealed by studies of other macromolecules emphasize this point and phototrophs provide a good example.

On the basis of 16S rRNA sequencing the cyanobacteria diverge from the same branch as the *Proteobacteria* and Gram-positive bacteria, while the green non-sulphur bacteria, such as the gliding organism *Chloroflexus aurantiacus*, represent a distinct and much earlier line which probably existed 3.5 billion years ago (Fig. 6.1). However, analysis of photoreaction centres implies the opposite relationship: the photosynthetic proteobacteria and *C. aurantiacus* have pheophytin quinone reaction centres whereas the cyanobacteria, the green sulphur bacteria and *Heliobacterium* (a strictly anaerobic Gram-negative organism, which represents a subdivision of the Gram-positive phylum and has the unique bacteriochlorophyll *g*) have the iron-sulphur type. *Chloroflexus aurantiacus* and the green sulphur organism *Chlorobium* have similar chlorosomes (light-harvesting surface structures) and yet analyses of the membrane-bound parts of the photosynthetic apparatus, including the reaction centres, all imply that *Chloroflexus* is an early-branching proteobacterial organism. The explanation for these anomalous observations is not yet known, but lateral genetic transfer has been postulated.

Another question concerns the evolutionary distances between the three domains, as different macromolecules give different answers. The archaea share several molecular features with the eukaryotes, such as the presence of the amino acid diphthamide in elongation factor (EF) 2, some tRNA genes containing introns (intervening sequences that do not contribute to the structure of the gene product), inhibition of protein synthesis by anisomycin but not chloramphenicol, and the complexity of DNA-dependent RNA polymerases. Also, trees based on sequence comparisons of RNA polymerase subunits, 5S rRNA, and ribosomal proteins show the archaea as having a smaller evolutionary distance to the eukaryotes than to the bacteria. In contrast, phylogenies based on 16S and 23S rRNA, and glyceraldehyde-3-phosphate dehydrogenase sequences place the archaea closer to the bacteria, and these differences of tree topology have not yet been explained.

Answers have been sought by simultaneous comparisons of pairs of genes such as EF-Tu and EF-G, the α and β subunits of ATPase, and lactate and malate dehydrogenases (LDH and MDH), which are believed to have diverged by duplication prior to the divergence of the three domains. The results generally support the view that the archaea are phylogenetically closer to the eukaryotes than to the bacteria but some anomalies remain owing to incomplete data: for example, the MDH sequence of the bacterium *Thermus flavus* has greater similarity to that of eukaryotes than that of bacteria.

Evolution of prokaryotes

Preliminary phylogenies based upon oligonucleotide cataloguing and, to a lesser extent, 5S rRNA and *c*-type cytochrome sequencing have now been revised and improved as the number of organisms characterized by full or partial sequencing of 16S rRNA and other molecules, including 23S rRNA genes, RNA polymerases, ribosomal protein genes and elongation factors, rapidly increases. Although the results of these studies confirm some groupings defined by phenotypic properties, such as the distinctness of the Gram-positive bacteria and the spirochaetes, the errors in many other assumptions are also being revealed. The Gram-negative bacteria, for example, do not represent a single genealogical group but comprise perhaps 10 separate divisions, which may be regarded as phyla or kingdoms. Sequence data do not support hypotheses suggesting that phototrophs and autotrophs arose from heterotrophs as these types do not appear in physiologically distinct groups but are intimately mixed within several divisions, and it seems unlikely that something so complex as the photosynthetic mechanism evolved more than once.

Figure 6.1 is a tree based on distance matrix analyses of 16S rRNA sequences and it shows the evolutionary relationships between the three domains or primary kingdoms and their divisions. The position of the tree's root is unknown, but as the domains have sequences in common it may be inferred that there existed a universal ancestor. This is presumed to have evolved sometime in the first billion years of the earth's existence and given rise to the photosynthetic bacteria found in 3.5 billion year old sedimentary deposits. It therefore appears that the three primitive domains diverged quite early and, as drastic changes would have been required to derive one domain from another, Woese has suggested that they evolved from a **progenote**, a rapidly evolving rudimentary organism that had primitive and inaccurate gene replication and translation systems and whose essential functions, although the most highly conserved, would show only general resemblances to those characteristic of modern cells.

The evolution of prokaryotes might be expected to show a relationship with the geochemical development of the earth, their phenotypes reflecting the nature of the environments in which they arose. Most of the archaea are strict anaerobes, many are thermophiles, and several are extreme thermophiles; perhaps the ancestor of the bacteria arose somewhat later, when the planet was cooler, as the thermophiles of this group show less extreme thermophily. None the less, green non-sulphur bacteria almost certainly existed more than 3.5 billion years ago, and it seems likely that the ancestor of the bacteria was an anaerobic, thermophilic autotroph or phototroph. More recently, aerobic phenotypes arose among the bacteria as oxygen accumulated in the earth's atmosphere from about 2.5 billion years ago.

Bacteria

On the basis of sequence distance analysis, characteristic (signature) sequences and structural features of 16S rRNA, the *Bacteria* comprise at least eleven divisions or phyla (Fig. 6.2) which will now be considered in turn.

The Proteobacteria

The class *Proteobacteria,* formerly called the purple bacteria, contains the purple non-sulphur and purple sulphur bacteria and most of the familiar Gram-negative genera. The long recognized need for splitting the genus *Pseudomonas* into several genera is emphasized by the presence of its members in three of the four natural subdivisions of the class. Gliding organisms also do not form a phylogenetically coherent group. Extensive DNA–rRNA hybridization studies by De Ley and colleagues at the

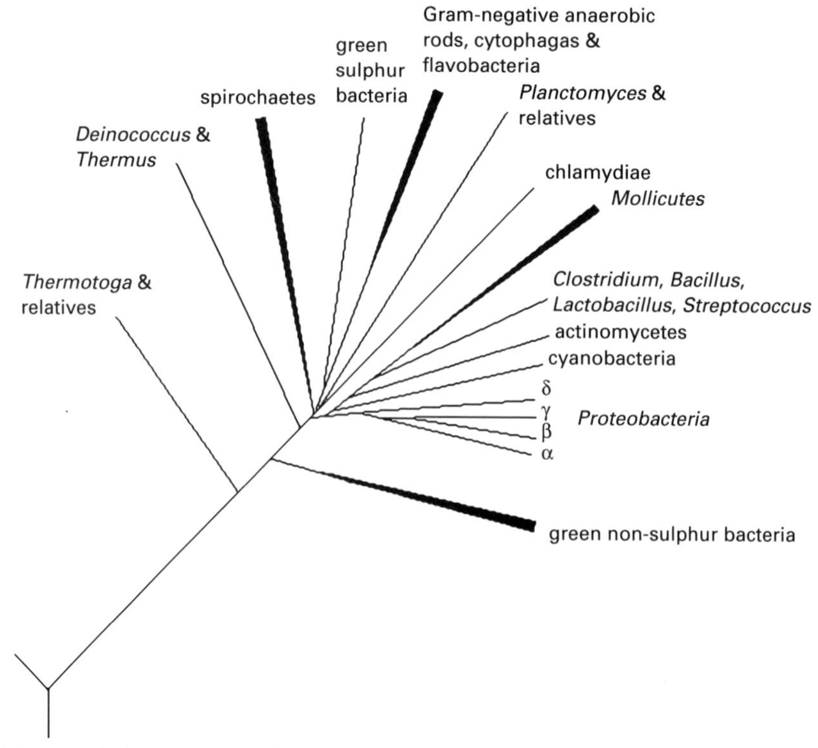

Fig. 6.2 Phylogenetic tree for *Bacteria* based on 16S rRNA sequence comparisons.

University of Gent have revealed the evolutionary relationships of the majority of genera within this class (Fig. 6.3) and their 'RNA superfamilies' correspond with the subclasses based on 16S rRNA sequencing.

The α Subclass (Superfamily IV). This contains most of the purple non-sulphur bacteria intermixed with a wide metabolic diversity of non-photosynthetic organisms. The latter are believed to have evolved from photosynthetic ancestors and aerobic metabolism appears to have evolved more than once. Further division correlates with phenotypic properties; in one group, which contains *Aquaspirillum* and *Azospirillum* species along with purple bacteria, helical organisms predominate. Asymmetric division or budding are common in a second group, which includes organisms such as *Nitrobacter* and *Rhodopseudomonas* species which oxidize or reduce nitrogenous compounds, respectively, suggesting connected metabolic evolutions; this group also contains "*Pseudomonas diminuta*" and "*P. vesicularis*".

It is of particular interest to find *Agrobacterium, Brucella, Rhizobium* and the rickettsial genera *Bartonella* and *Rochalimaea,* which contain intracellular or intimate extracellular parasites of eukaryotic cells together in this, the subclass that yielded an intracellular symbiont which gave rise to mitochondria. Other rickettsias, including *Anaplasma, Cowdria, Ehrlichia* and *Rickettsia,* form a distinct subline. Prosthecate organisms such as *Caulobacter* and *Prosthecomicrobium* also belong to the α subclass.

The β Subclass (Superfamily III). The remaining genera of purple non-sulphur bacteria, *Rhodocyclus* and *Rubrivivax,* and many non-photosynthetic genera are found here. *Rhodocyclus* species are not only distinguished from α subclass photosynthetic species by rRNA sequences, but also on account of having a smaller subunit type of cytochrome *c* and a different photochemical reaction centre structure. Deep and dense branching, sometimes of uncertain order, makes splitting of the subclass unclear, but at least two subgroups may be recognized. One contains *Alcaligenes, Bordetella,* "*Pseudomonas*" species (now in *Comamonas, Acidovorax* and *Hydrogenophaga,* family *Comamonadaceae*), *Rubrivivax, Rhodocyclus* (possibly representing a distinct and separate line), and species of *Nitrosomonas* and *Nitrosolobus* which oxidize ammonia. The other includes the family *Neisseriaceae* (*Eikenella, Kingella, Neisseria* and *Simonsiella,* the last of which comprises gliding organisms), and *Vitreoscilla stercoraria* (another gliding organism), *Aquaspirillum, Chromobacterium* and *Iodobacter* which branch progressively deeper.

The γ Subclass. This is closely associated with the β subclass and contains several deeply branching subgroups. One contains the purple sulphur bacteria of *Chromatiaceae* and *Ectothiorhodospiraceae,* and a second (superfamilies I and II) contains a wide variety of non-photosynthetic groups including the enterobacteria, vibrios and fluorescent (true) pseudomonads,

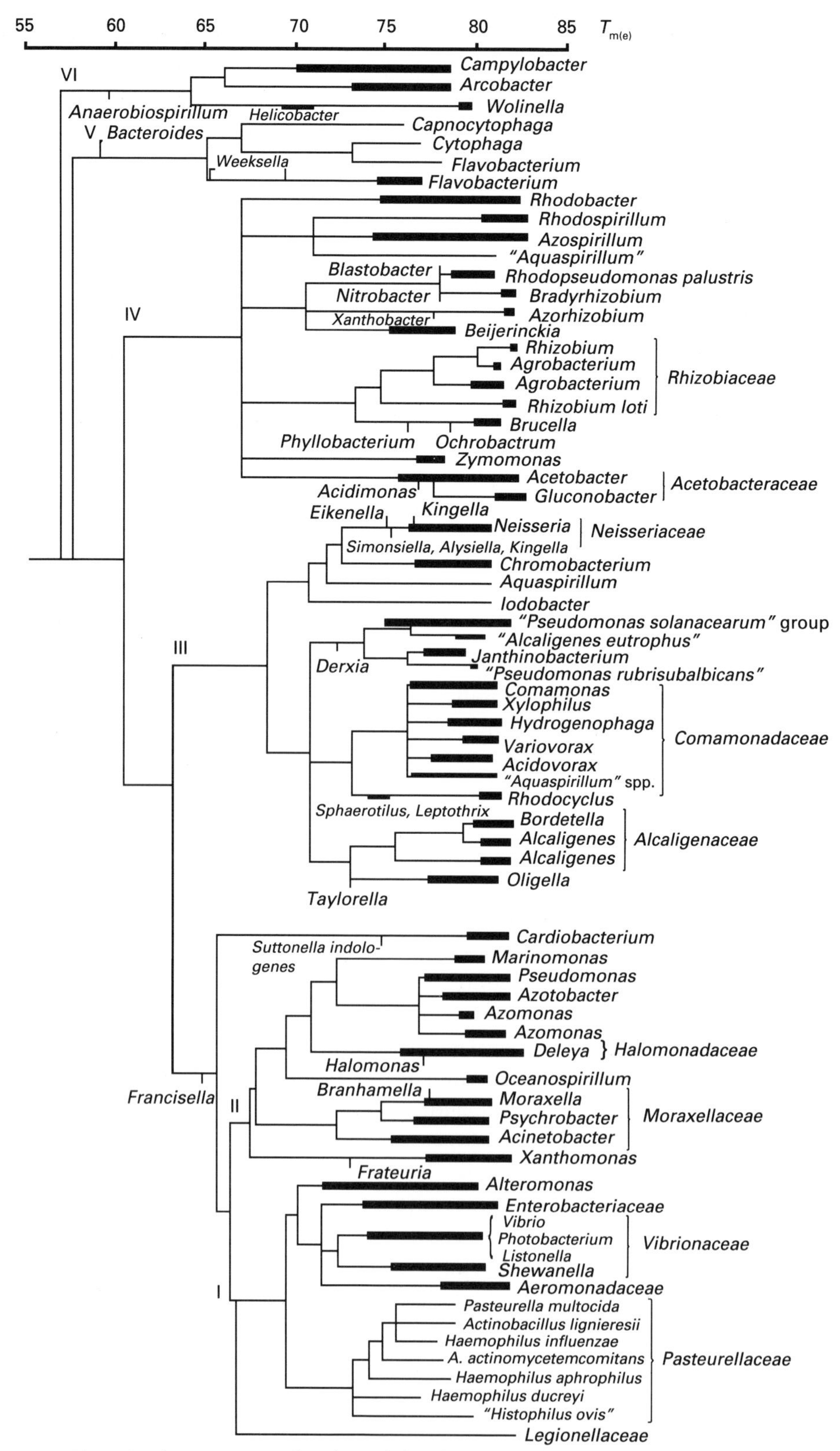

Fig. 6.3 The RNA superfamilies of the class *Proteobacteria* as revealed by DNA–rRNA hybridization studies carried out at the University of Gent.

Acinetobacter, Aeromonas, Oceanospirillum, Pasteurella, Xanthomonas and gliding organisms such as *Alysiella, Beggiatoa* and *Leucothrix*. Unexpectedly, *Legionella* and the rickettsial genus *Coxiella*, which have intracellular parasitism in common, show some relationship and appear to form a third subgroup in which the aphid endosymbiont *Buchnera* represents a distinct line. The relationship of another rickettsia, *Wolbachia*, to this subgroup is uncertain.

The δ Subclass. Three quite different phenotypes are found in this subclass. The first two, bdellovibrios, which are intracellular parasites of bacteria, and gliding bacteria of the order *Myxobacterales*, both appear to form distinct groups. The third type, sulphur and sulphate reducing bacteria such as *Desulfobacter* and *Desulfovibrio* and their relatives, form several groups. The relationship between the three types is not clear but the common ancestor was probably a sulphate reducing anaerobe. Energy yielding metabolisms based upon the reduction of sulphur compounds are also found among Gram-positive bacteria and the archaea.

Campylobacteraceae and Relatives (Superfamily VI). Campylobacter, Wolinella, the misnamed *Bacteroides* species *"B. gracilis"* and *"B. ureolyticus"*, and the sulphur oxidizer *Thiovulum* appear to represent a further subclass of the *Proteobacteria*. Other Gram-negative sulphur and iron oxidizers such as *Acidophilium, Thiobacillus* and *Thiothrix* are found in the α, β and γ subclasses.

The Gram-positive bacteria

Despite the title, the anaerobic organisms which comprise two of this division's four recognized subdivisions do not have Gram-positive cell walls, but they are poorly characterized at present. One includes the genera *Megasphaera, Selenomonas* and *Sporomusa*, the last of which contains endospore formers, and the other has but a single representative, the anoxygenic phototroph *Heliobacterium chlorum*, whose unique bacteriochlorophyll *g* shows some structural similarity to chlorophyll *a* of the cyanobacteria, and whose photoreaction centre is of the type found in the green sulphur bacteria and the cyanobacteria.

The other two branches correlate with G+C contents, one containing species with less than 50 mol% G+C, and the other 55 mol% G+C or more, but, as currently defined, some genera such as *Bacillus* span the range. The lower mol% G+C branch is deeply divided and seems to be the more ancient group, and its aerobic and aerotolerant members may have evolved from an anaerobic spore-forming ancestor in parallel with changes in the earth's atmosphere. This branch contains the two endospore-forming genera *Bacillus* and *Clostridium*, and *Lactobacillus, Staphylococcus* and *Streptococcus*, along with the mollicutes. The mollicutes do not have cell walls, and form a broad, distinctive, rapidly evolving group whose 16S rRNAs show changes in some of the most highly conserved positions.

The higher mol% G+C lineage has shallower branchings and appears to be less ancient than the lower mol% G+C one. It contains the actinomycetes, most of which show pleomorphism or branching growth and are aerobic; the anaerobic genera *Bifidobacterium* and *Propionibacterium* are the deepest branching members. Other genera include *Actinomyces*, *Arthrobacter*, to which *Micrococcus* is closely related, *Corynebacterium*, *Mycobacterium* and *Streptomyces*. Sulphate-reducing endospore-formers, placed in the heterogeneous genus *Desulfotomaculum*, also belong in this division but their precise affiliations are not known.

Cyanobacteria

There is a great diversity of cyanobacteria and they are widely distributed, but they form a distinct phylogenetic group and all produce bacteriochlorophyll *a* and carry out oxygenic photosynthesis. Green chloroplasts are found within the division, thus confirming their endosymbiotic origins, and *Prochloron*, a group of bacteriochlorophyll *a* and *b* producing organisms, which are extracellular symbionts of marine invertebrates, also shows some relationship.

This division appeared quite late compared with most other bacteria and may be related to the Gram-positive bacteria (see Molecular Clocks above). Also, the different cyanobacterial types diverged over a relatively short period, perhaps because they were the first in a new niche by using water as an electron donor for photosynthesis. Groupings based upon phenotypic characters such as presence of heterocysts, branching growth and production of baeocytes (small coccoid reproductive cells), show some correlation with the phylogeny, but unicellular and filamentous types are dispersed throughout the tree.

The spirochaetes

This distinct division, which comprises the helically coiled bacteria with periplasmic flagella, is deeply branched and probably very ancient; indeed, on the basis of 16S rRNA sequences some species of this phylum are as similar to enterobacteria as they are to other spirochaetes. *Leptonema* and *Leptospira* form one subdivision and this is deeply divided from the other, which contains several well-separated sublines. These include the genus *Serpulina* (formerly in *Treponema*) which branches deeply from a main group composed of two sublines; one contains *Treponema*, and the other contains the majority of *Spirochaeta* species and *Borrelia*, the latter being a tight cluster branching early from the former line.

Gram-negative anaerobic rods, cytophagas and flavobacteria

Unrecognizable on phenotypic grounds, this odd grouping corresponds to RNA superfamily V and comprises two major subdivisions: one is composed of the Gram-negative anaerobic rods such as the genera *Bacteroides* and *Fusobacterium*, but the other consists of aerobes and contains a mixture of *Flavobacterium* (*sensu stricto*) species with genera of gliding bacteria such as *Cytophaga*, *Flexibacter*, *Saprospira* and *Sporocytophaga*. However, as *Flexibacter* species are not closely related to one another but are distributed about the phylum, and there is evidence for a third and intermediate subdivision containing anaerobic, flexible, gliding organisms, it has been inferred that gliding motility may have been an ancestral property of the phylum. The family *Spirosomaceae*, whose members have ring-like morphologies, also belong to this phylum.

Green sulphur bacteria

Only a few strains of *Chlorobium* and *Chloroherpeton* have been characterized, but these anoxygenic phototrophic organisms appear to form a tight and shallow-branching group with no close photosynthetic or non-photosynthetic relatives. However, a distant relationship with the *Bacteroides* group has been inferred from 16S rRNA signature and structural similarities, and 23S rRNA sequences. Although members of *Chloroherpeton* and the green non-sulphur genus *Chloroflexus* both show gliding motility and resemblances in chlorosomes and chlorophyll type, their photoreaction centres differ and their rRNAs are not related.

Chlamydiae

This small, distinct division contains just four species, the close relatives *Chlamydia psittaci* and *C. trachomatis* and the more distant *C. pecorum* and *C. pneumoniae*, all of which are parasites of homotherms and whose cell walls lack peptidoglycan (see Chapter 2). Ribosomal RNA signature and structural similarities suggest a remote relationship with another division of peptidoglycan-less organisms, *Planctomyces* and their relatives.

Planctomyces and relatives

This lineage evolved rapidly and has a distinctive 16S rRNA signature. Its members, *Isosphaera* and *Planctomyces*, are budding, non-prosthecate organisms, and they possess fewer of the highly conserved rRNA oligonucleotides than any other division.

Deinococcus and relatives

Although Gram-positive and superficially similar to *Micrococcus*, the radiation resistant organisms forming *Deinococcus* resemble Gram-negative bacteria in their lipid profiles and possession of outer membranes containing protein and lipid. They represent one of two subdivisions of a unique and early branching phylogenetic group, members of which have the same peptidoglycan type; the other subdivision contains the widespread genus of thermophilic rods, *Thermus*.

The green non-sulphur bacteria and relatives

This division includes three genera of flexible, gliding bacteria, *Chloroflexus*, *Heliothrix* and *Herpetosiphon*, the first two of which contain oxygenic phototrophs, and a genus of non-motile, pleomorphic thermophiles, *Thermomicrobium*, which probably belongs to a separate subdivision. The lineage is an ancient one which branched off well before the period of intense bacterial divergence, and its members share a distinctive rRNA signature and structure. Chemotaxonomic studies have also revealed some phenotypic similarities between the subdivisions. Analyses of photosynthetic apparatus suggest a relationship between *Chloroflexus* and the *Proteobacteria*, an unexplained anomaly (see Molecular Clocks above).

Thermotoga and relatives, and other divisions

Members of this genus of extremely thermophilic anaerobes, with unique lipids and sheath-like structures, have been isolated from marine waters and springs. Along with the related genera *Fervidobacterium* and *Thermosipho* they form the earliest branch of the bacteria recognized to date and probably represent a much larger group of organisms awaiting discovery. As further bacteria are found and characterized, it is likely that new divisions will be revealed; one may be represented by the prosthecate organism *Verrucomicrobium spinosum*, and another by species of *Thermodesulfotobacterium*, a genus of thermophilic, sulphur reducing rods which are the only bacteria with ether-linked lipids and do not appear to belong to any of the divisions recognized so far.

Archaea

Before the recognition of this second prokaryote kingdom or domain the peculiarities of organisms now included in it were thought to represent adaptations of eubacteria to special, often extreme, environmental niches. The domain is broad and contains organisms with many metabolic differences, a diversity of cell wall types, a wide range of shapes, sizes and pigmentation, and multiplication by a variety of mechanisms. Chemotaxonomic and genomic analyses, however, have revealed some shared distinguishing features; these include the presence of unique ether-linked lipids, cell walls lacking muramic acid or the usual form of peptidoglycan, and characteristic ribosomal components and structure.

The archaea do not have a common cell wall polymer as the bacteria do. It appears that effective polymers only evolved after considerable branching and physiological diversification among organisms had occurred, and these polymers are therefore valuable taxonomic markers. Two of the commonest archaeal cell wall types are associated with positive and negative Gram-staining reactions. The former has a layer of **pseudomurein** (differing from the bacterial polymer in the substitution of *N*-acetyltalosaminuronic acid for *N*-acetylmuramic acid and having L-amino acid tetrapeptide

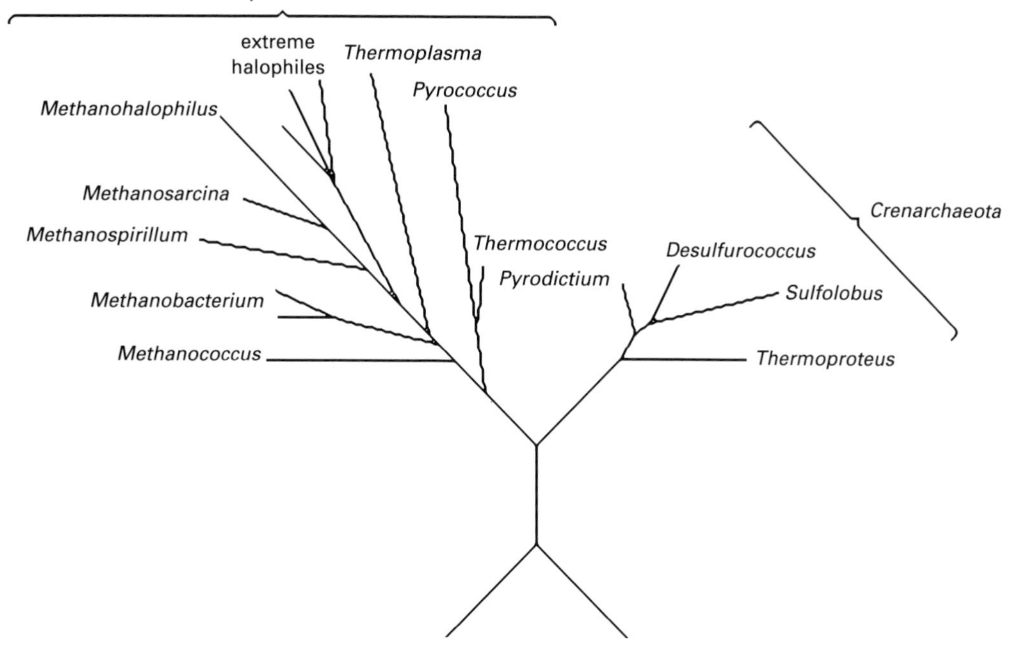

Fig. 6.4 Phylogenetic tree for the *Archaea* based on 16S rRNA sequence comparisons.

crosslinks) in the *Methanobacteriales*, or an acidic heteropolysaccharide which may be sulphated or not (examples being *Halococcus* and *Methanosarcina* respectively). The latter has a surface (S) layer of hexagonally or tetragonally arranged protein or glycoprotein units as the only cell wall component; such cell walls are easily disrupted and are found in all the extreme thermophiles and in many halophiles and methanogens. Antibody probing of the S layer by indirect immunofluorescence has proved valuable in the taxonomic characterization of the methanogens.

Although some groups show evidence of relatively rapid development, archaeal evolution appears to be relatively slow. The 16S rRNAs of the other two domains are more similar to those of the archaea than they are to each other, suggesting that the archaea are closer to the postulated universal ancestor and are the more primitive of the two prokaryote domains. Also, on the basis of some molecular similarities, the archaea seem, on balance, to be closer to the eukaryotes than to the eubacteria (see Molecular Clocks above).

There are two main archaeal divisions (Fig. 6.4), and they have distinctive DNA-dependent RNA polymerases and show differing amounts of tRNA modification. It has been suggested that the divisions be regarded as kingdoms *Crenarchaeota* and *Euryarchaeota* of the *Archaea*. Some authors recognize the *Thermococcales* as a third major division. As both kingdoms include deeply branching groups of slowly evolving, extremely thermophilic sulphur reducing organisms, it has been inferred that the archaeal ancestor was also a sulphur reducing thermophile.

Crenarchaeota — the extreme thermophiles

The name *Crenarchaeota* is derived from the Greek for spring or fount, to indicate the phenotypic similarity of these organisms to the supposed archaeal ancestor. This division or kingdom contains a variety of organisms isolated from marine and terrestrial volcanic vent environments including 'black smoker' waters, flanges (horizontal outcrops of metal sulphide mounds which trap rising hydrothermal fluids), acidic hot springs, solfataric waters (which emit SO_2), boiling mud holes and other geothermally heated niches. The submarine hydrothermal vent environments are particularly interesting as their primary producing floras comprise anaerobic and aerobic chemolithoautotrophs rather than phototrophs.

All of the organisms in this kingdom are quite recent discoveries, most having been described since 1980, and because of their apparently important part in evolution they have attracted great interest. New species, genera and groups are regularly being discovered, but as descriptions are often based on very few strains, the boundaries of groups at the genus and family levels, and so affiliations of higher taxa, can be hard to define. Consequently the taxonomy of these organisms is less advanced than that of the other archaeal kingdom and remains in a state of flux. Most classification has been based on 16S rRNA sequencing, with support being given by mol% G+C and morphological and physiological characteristics. At the species level DNA relatedness, chemotaxonomic methods such as lipid patterns, and antibody probing of S layers and other proteins are used.

The organisms range from the deeply branching and presumably most ancient lines of strictly anaerobic, extremely thermophilic hydrogen-sulphur autotrophs, most of which have been assigned to the order *Thermoproteales*, to the aerobic and facultatively aerobic extreme thermoacidophiles accommodated in the *Sulfolobales*. Genera include: *Pyrodictium*, which contains slowly evolving, disk-shaped hydrogen-sulphur autotrophs and heterotrophs growing at temperatures up to 110°C and represents a distinct line; rods or filaments of the *Thermoproteaceae*, capable of sulphur respiration (energy-yielding metabolism in which elemental sulphur is used, anaerobically, as the terminal electron acceptor) of organic matter with growth temperatures between 80 and 100°C (*Thermoproteus* and the acidophile *Thermofilum*); cocci capable of sulphur respiration and fermentation (*Desulfurococcus*) and heterotrophy in the presence of sulphur (*Staphylothermus*); *Hyperthermus*, an anaerobic, peptide-fermenting genus which grows at temperatures up to 108°C and uses H_2S production as an accessory energy source; facultatively anaerobic thermoacidophiles capable of lithotrophic growth by oxidation or reduction of sulphur (*Acidianus*) and aerobic thermoacidophiles which oxidize sulphur compounds or organic material and grow between 50 and 87°C (the rapidly evolving genus *Sulfolobus*), both of which are placed in the family *Sulfolobaceae*.

Other genera, of unknown affiliation, are: *Metallosphaera*, thermoacidophilic metal-mobilizing organisms that may belong to the *Sulfobolaceae*; *Pyrobaculum*, which are extremely thermophilic, facultative chemolithoautotrophs and obligate heterotrophs; *Sulfosphaerellus* and *Stygiolobus*, both obligate chemolithoautotrophs; an unnamed, vitamin requiring chemoorganotroph from a flange, growing at 110°C and *Caldococcus*, which are peptide fermenting, extremely thermophilic sulphur reducers.

Euryarchaeota — the methanogens and halophiles

This kingdom was named for the great diversity seen within it: there are three subdivisions of methanogens, one of extreme halophiles, and three other lineages whose affiliations are uncertain. There seem to have been several periods of rapid evolution and these can be traced from methanogens with very limited ranges of energy sources, through those that are less restricted, to halophilic methylotrophs and finally to the extremely halophilic aerobes.

The methanogens are strict anaerobes which possess unusual enzyme cofactors and form a physiologically specialized group. They obtain energy by oxidizing hydrogen or simple organic compounds and use the electrons generated to reduce carbon dioxide to methane. They occur in aquatic sediments, salt marshes, intestinal tracts and the rumen, sewage sludges, and oil fields, and range from mesophiles to extreme thermophiles. They have long attracted interest as the only microorganisms producing a hydrocarbon as the major metabolic product and as such they are of great environmental and economic importance.

Before the recognition of the archaea the methanogens used to be classified as a physiologically defined family, and it is of evolutionary interest that many strains can reduce sulphur, although this may not support growth. In taxonomic terms they are the best studied and most rapidly expanding group of archaea. This, however, has led to some difficulties, as descriptions of species have often been based on few strains and a very limited range of characters, so minimal standards for methanogen taxonomy have been suggested. In addition to nucleic acid analyses, characters widely used include antibody probing of S layers, protein electrophoresis, polar lipid patterns, polyamine patterns and determinations of catabolic substrates, growth factors, and pH and temperature requirements using gas chromatographic measurements of methane production.

Methanococcales

Methanococcus comprises obligately methanogenic, halophilic and mesophilic to extremely thermophilic, irregular cocci that have protein cell walls and use only hydrogen and formate as electron donors; it is not a very homogeneous group and may represent more than one genus. Relatively few strains have been characterized, and the values of the known distinctive phenotypic characters are uncertain and become more so as further strains are studied. The establishment of the separate order is therefore largely based on 16S rRNA analysis. Identification used to be based on morphology and growth characteristics, and closer study of two species resulted in their transfer to *Methanosarcina*.

Methanobacteriales

Morphologies vary from cocci to filaments, but cell walls all contain pseudomurein. The organisms are strictly anaerobic, and are highly specialized in not catabolizing organic matter other than carbon monoxide and formate which, with hydrogen, are the only energy sources. The order was divided on the basis of rRNA cataloguing. *Methanobacteriaceae* contains two genera which do not grow above 70°C, namely *Methanobacterium* (long, Gram-positive or Gram-variable rods), and *Methanobrevibacter* (Gram-positive coccoid rods which require certain vitamins). The single-member family *Methanothermaceae* contains Gram-positive, sulphur reducing, chemolithotrophic rods which grow at 60°C to about 95°C, and have an outer protein layer to the cell wall. Cell wall composition, immunological fingerprinting and 16S rRNA cataloguing indicate that *Methanosphaera*, a genus of methanol reducing cocci, is related to this order.

Methanomicrobiales

Members of this order have proteinaceous cell walls and no pseudomurein, span a wide range of morphologies, and catabolize methyl groups. There are three phylogenetically defined families: *Methanomicrobiaceae*, containing rods requiring acetate (and sometimes other organics) as a nutrient (*Methanolacinia* and *Methanomicrobium*), helical, sheathed rods (*Methanospirillum*), halotolerant, irregular cocci (*Methanogenium*, a heterogeneous genus from which species have been removed to *Methanocorpusculum* and *Methanoculleus*); *Methanosarcinaceae*, comprising cocci which typically form aggregates or cysts and may contain gas vesicles (*Methanosarcina*), irregular cocci with or without internal membrane structures (*Methanolobus* and *Methanococcoides* respectively), sheathed rods catabolizing only acetate (*Methanosaeta* and *Methanothrix*), and halophilic cocci (*Methanohalophilus*); and *Methanocorpusculaceae*.

The affiliation of *Methanoplanus*, a genus of plate-shaped, chemolithotrophic organisms, one species of which is an endosymbiont of a marine ciliate, is unclear; 16S rRNA hybridization and oligonucleotide cataloguing respectively indicated family status and a relationship with *Methanomicrobiaceae*. A novel genus of uncertain position is *Methanopyrus*, whose members are chemolithoautotrophs growing at temperatures up to 110°C.

Extreme halophiles

These aerobic and facultatively anaerobic, chemo-organotrophic organisms are distinctly pleomorphic, ranging from rods (*Heliobacterium* and the alkaliphilic genus *Natronobacterium*) through cocci (*Halococcus* and the alkaliphilic genus *Natronococcus*) to pleomorphic cells forming rods, rectangles, triangles and disks (*Haloarcula* and *Haloferax*), but they form a relatively tight phylogenetic group, that is not divisible into families, which is believed to have arisen from the *Methanomicrobiales*. They are also phenotypically distinctive as they require NaCl concentrations between 8% and saturation, produce carotenoid pigments, and can, in some cases, harvest light energy using the pigment bacteriorhodopsin. They are ubiquitous in very salty environments such as salt lakes, soda lakes, salterns, and in solar salts and heavily salted proteinaceous matter like fish and hides.

Classification within the group has been confused by the placement of unrelated organisms in the same taxa over the years, so that strains having only their names in common were used as taxonomic markers, and comparisons of several studies which omitted type strains are difficult. Polyphasic studies including immunological tests are helping to resolve the picture, with polar glycolipid patterns being particularly helpful in defining the non-alkaliphilic groups, and quantitative differences in diether lipids being of some value with the alkaliphiles. None the less, the generic affiliations or synonymies of some organisms are still not clear, and several species of uncertain taxonomic position remain attached to *Halobacterium* for example, for while nucleic acid homologies and chemotaxonomic features are gradually allowing resolution of groups, these groups are not always discernible by numerical taxonomy.

Thermoplasmas

These cell wall-deficient, obligate thermoacidophiles are facultatively anaerobic by sulphur respiration, and have been isolated from coal waste and solfatara fields. They appear to belong to the methanogen division and represent a rapidly evolving branch of uncertain position. Interestingly, the mollicutes also form a rapidly evolving line.

Thermococcales

This order presently comprises a single family containing the two genera *Pyrococcus* and *Thermococcus* and represents a deeply branched and slowly evolving line of extremely thermophilic and strictly anaerobic organisms that utilize carbon sources by sulphur respiration (and in *P. furiosus* by fermentation also). Serology of DNA-dependent RNA polymerase indicates that *Pyrococcus* belongs to the *Thermococcaceae*, while DNA–rRNA hybridizations show that it represents a long branch compared with the short and presumably more slowly evolving one of *Thermococcus celer*. Data from rRNA sequencing and DNA–rRNA hybridizations suggest that the family line is about equidistant between the methanogen-halophile and extreme thermophile divisions, but much depends on where the archaeal root lies; Woese included the thermococci in the former kingdom, the *Euryarchaeota*, and the many 16S rRNA signature sequences *Thermococcus* shares with the methanogens, the organization of their rRNA genes, and dendrograms based on comparisons of EF-1α and S10 ribosomal protein sequences support this allocation.

Sulphate reducers

Archaeoglobus fulgidus and *A. profundus* are species of extremely thermophilic sulphate reducing cocci isolated from volcanic vent environments. They show fluorescence at 420 nm, which is characteristic of the methanogens, and Woese included the genus in the *Euryarchaeota*, but the RNA polymerase of *A. fulgidus* is distinct from those of the main archaeal divisions, suggesting that it might belong to a third major division; by any standards the genus appears to represent a new order.

Conclusion

The foregoing outline of prokaryote evolution is mainly based upon 16S rRNA analyses, but other genomic and chemotaxonomic studies such as enzyme comparisons tend to support it. In any case it is clear that many of the phenotypic characters useful for identification, such as shape, motility and photosynthesis, are of little value in phylogenetic studies.

As full 16S rRNA sequence and other molecular data for well-established and newly described taxa accumulate, branching orders will be resolved and further divisions and subdivisions will no doubt come to light. Bacterial phylogeny is a rapidly evolving discipline.

7 The spirochaetes

I have also seen a sort of animalcules that had the figure of our river-eels: these were in very great plenty, and so small withal, that I deemed 500 or 600 of 'em laid out end to end would not reach to the length of a full-grown eel such as there are in vinegar. These had a very nimble motion, and bent their bodies serpent-wise, and shot through the stuff as quick as a pike does through the water.

Antonie van Leeuwenhoek, 1681, Letter to the Royal Society

Introduction

The special nature of the spirochaetes has long been recognized. In 1835 Ehrenberg proposed the genus *Spirochaeta* for spirally wound, flexible filaments and in 1917 Buchanan, noting that they were 'protozoan like in many characters', accommodated such organisms in the order *Spirochaetales*. Since then phenotypic observations, especially by light and electron microscopy, have added to the definition of the taxon, further isolates have conformed, and together they have emphasized the group's uniqueness. It is a rare example of a taxon delineated by phenotypic characters that is also phylogenetically distinct according to rRNA sequencing studies.

All members of the order are flexuous, helically shaped, Gram-negative, chemoheterotrophic organisms which differ from other prokaryotes by having between two and more than one hundred periplasmic flagella permanently wound around the cell; they are inserted at the ends of the cell, overlap in the middle, and are enclosed within an outer sheath (Fig. 7.1). Spirochaetes can flex, rotate and creep on surfaces, and are motile even in quite viscous environments. Their mechanisms of locomotion have not been explained, but must be unlike those of other bacteria.

Notwithstanding these similarities, the spirochaetes form a very heterogeneous group with a mol% G+C range of 25 to 65, and although 16S rRNA signature sequences show spirochaetes to be of **monophyletic** origin (representing a single evolutionary line), leptospires show higher sequence similarities to *Escherichia coli* than to some treponemes. Their carbon and energy sources include carbohydrates, amino acids, and long-chain fatty

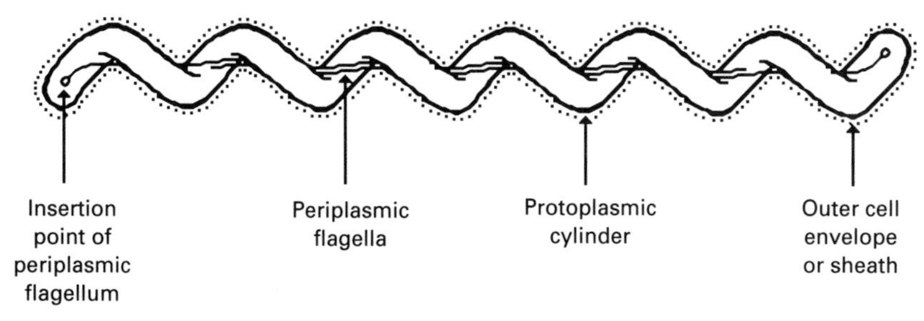

Fig. 7.1 Schematic diagram of a spirochaete.

acids and alcohols, and they have gaseous requirements varying from aerobic through microaerophilic and facultatively anaerobic to strictly anaerobic. Their modes of existence are extremely diverse and range from saprophytism in fresh and marine waters, sediments and soil to obligate parasitism as commensals or pathogens in wood-eating insects, molluscs, blood-sucking arthropods and mammals.

Classification

On the bases of morphological, physiological and chemotaxonomic characters the spirochaetes are divisible into two families, a division which is supported by 16S rRNA sequencing studies; members of the *Spirochaetaceae* use carbohydrates and amino acids as carbon and energy sources, are obligately or facultatively anaerobic or microaerophilic, have L-ornithine as their peptidoglycan diaminoacid, and rarely have hooked ends, whereas *Leptospiraceae* members use long-chained fatty acids and fatty alcohols as carbon and energy sources, are aerobic, have diaminopimelic acid in their peptidoglycan, and usually have hooked ends.

Spirochaetaceae

The classification of this family is hampered by missing data owing to difficulties of cultivation, and none of the type species of three of the genera, *Cristispira*, *Spirochaeta* and *Treponema*, have been grown in pure culture. The genera are distinguished by the morphologies and habitats of their members and by the few physiological characters that have been determined for representative numbers of species, and from limited genomic analysis it is clear that *Spirochaeta* and *Treponema* are heterogeneous.

Spirochaeta. This genus contains free-living obligately anaerobic and facultatively anaerobic spirochaetes which use carbohydrates as energy and carbon sources and are capable of the *de novo* synthesis of cellular fatty acids. They are found in freshwater and marine environments, particularly those containing H_2S, such as the waters and sediments of ponds, swamps and rivers, intertidal muds and thermal springs. The six cultivable species are morphologically similar, having slender cells with two periplasmic flagella, and their genomes show a range of 51–65 mol% G+C.

They have been divided into two groups on the basis of their relations to molecular oxygen, and the obligate anaerobes lie in the lower part and the facultative anaerobes in the higher part of the mol% G+C range. However, the findings of 16S rRNA sequencing studies do not correlate with this division; the obligate anaerobe *S. stenostrepta* and the facultative anaerobe *S. zuelzerae* are more closely related to each other, and to *Treponema denticola* and *T. phagedenis*, than they are to the obligate anaerobe *S. litoralis* (Fig. 7.2) and the facultative anaerobes *S. halophila* and *S. aurantia*, and they are probably misclassified. Any revision of the genus must await the cultivation and characterization of *S. plicatilis*,

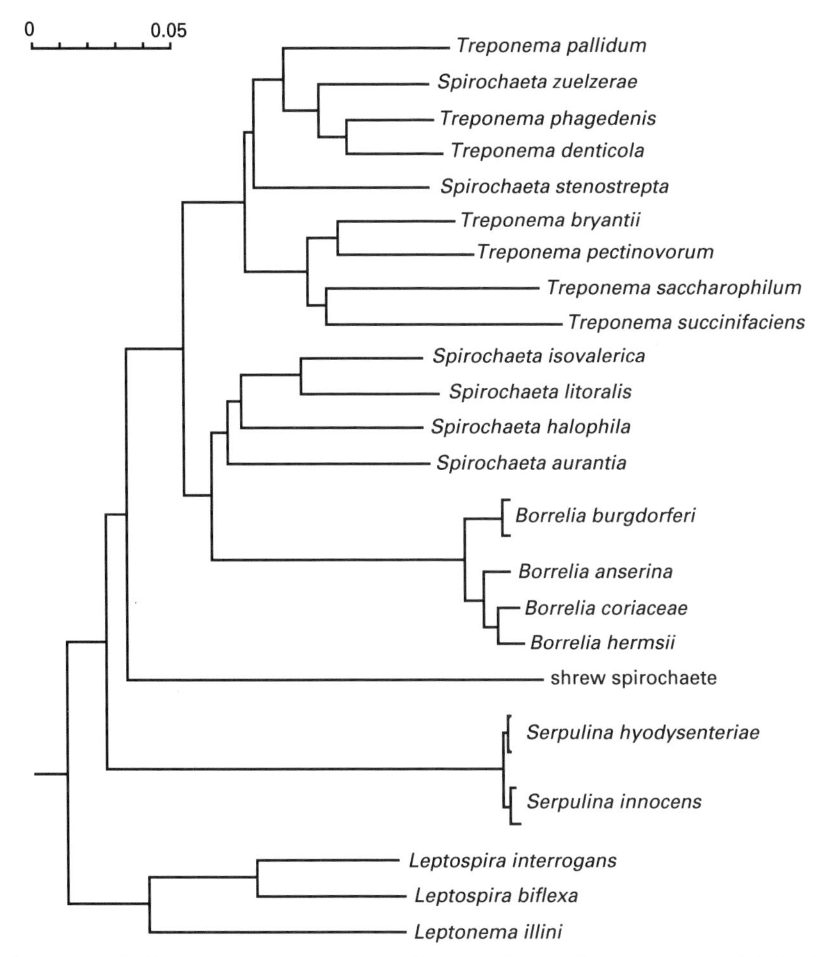

Fig. 7.2 A phylogenetic tree of the spirochaetes, adapted from Paster *et al.* (1991). Bar represents a 5% difference in nucleotide sequences.

the type species (whose cells, incidentally, are much larger than those of the other species), and there are, without a doubt, many other species of free-living, anaerobic and facultatively anaerobic spirochaetes still to be described.

Treponema. The 14 species of this genus, three of which are non-cultivable, are host-associated, include anaerobes and microaerophiles, and span a range of 36–54 mol% G+C. Some of the cultivable species use carbohydrates as sources of energy and carbon whereas others use amino acids. They also vary in cell size and products of metabolism, and although all require fatty acids they differ in the types needed.

Again, revision of the genus is hindered by the non-cultivability of its type species *T. pallidum*. Splitting into several genera according to mol% G+C and phenotypic data, along the lines indicated in Table 7.1, is only partly supported by genomic data, and antigens shared between members of the groups have been detected. The representative species *T. denticola*, *T. pallidum* and *T. phagedenis* show little relatedness to one another in

Table 7.1 Subdivisions of *Treponema*

	Non-cultivable	Cultivable		
		Non-fermentative	Fermentative	
Species	*T. pallidum* subspecies*	*T. denticola*	*T. bryantii*	*S. hyodysenteriae*†
		T. minutum	*T. phagedenis*	*S. innocens*
		T. refringens	*T. succinifaciens*	
		T. scoliodontum		
		T. vincentii		
Cell width (μm)	0.13–0.15	0.15–0.25	0.20–0.30	0.36–0.38
mol% G+C range	52.4–53.7	37.0–43.0	36.0–39.0	24.7–29.7

*No data available for the other non-cultivable species *T. carateum* and *T. paraluiscuniculi.*
†*Serpulina*, formerly classified as *Treponema.*

DNA–DNA hybridizations, but *T. refringens* also shows very little relatedness to *T. denticola* and to most of the other species and strains.

Furthermore, 16S rRNA sequence analyses suggest that *T. denticola, T. phagedenis* and *T. pallidum* show quite a close evolutionary relationship, as do *T. bryantii, T. pectinovorum, T. saccharophilum* and *T. succinifaciens*, yet these treponemes are as close to certain species of *Spirochaeta* and strains tentatively assigned to that genus (which are probably true treponemes) as they are to each other. *Treponema hyodysenteriae* and *T. innocens* have a low mol% G+C range and appear to represent an evolutionary branch peripheral to the main *Spirochaeta–Treponema–Borrelia* cluster. They have been placed in a new genus, *Serpulina*, following DNA hybridizations, DNA restriction endonuclease analysis, PAGE of cell proteins and 16S rRNA sequencing. Further distant is a representative of a group of spirochaetes isolated from shrews and mice during studies on Lyme disease; they are morphologically and serologically distinct from *Borrelia, Leptospira, Spirochaeta* and *Treponema* and probably represent a new genus.

All this actual and implied splitting is partly balanced by *T. pertenue* becoming a subspecies of *T. pallidum* on account of these species showing 100% relatedness in DNA–DNA hybridizations; another subspecies has been proposed for the *T. pallidum* variant that causes non-venereal syphilis.

Borrelia. Members of this genus are tick- or louse-borne pathogens of mammals and birds. They are distinguished from other parasitic spirochaetes by their loose, irregular coiling and stainability, but not many have been grown *in vitro*, and few physiological, biochemical and genomic data are available. The 22 species have been classified largely according to their arthropod vectors and the diseases they cause; for example, the three species *B. hermsii*, *B. parkeri* and *B. turicatae* cause tick-borne relapsing fever in western parts of North America and show complete specificity for their respective tick vectors *Ornithodoros hermsi*, *O. parkeri* and *O. turicata*. Such specificity is not absolute for all members of the genus, and patterns of infectiveness for different vectors, transmissibility by them, and infectivities for experimental animals have also been used taxonomically.

The G+C range for the species examined to date is 27–32 mol%, and this is well outside the ranges for most of the other spirochaetes. More promising are DNA–DNA hybridization studies, the findings of which may be interpreted alongside ecological and cultural characters: for instance, *B. parkeri* and *B. turicatae* showed 77% and 86% relatedness to *B. hermsii* respectively and so these three might be regarded as one species. *Borrelia burgdorferi*, the agent of Lyme disease, *B. coriaceae*, the putative agent of epizootic bovine abortion, and *B. anserina*, which is pathogenic for birds, showed DNA–DNA relatedness of 30–44%, 44–50% and 53–63%, respectively, to *B. hermsii* and so, on this basis, the genus appears to be a natural one.

Study of further isolates of *B. burgdorferi* showed this species to be phenotypically and genotypically heterogeneous. Following the identification of three DNA homology groups, and associated specific patterns revealed by rRNA gene restriction endonuclease analysis, protein electrophoresis, and reactions with monoclonal antibodies, the new species *B. garinii* was proposed.

In 16S rRNA studies a few strains of *Borrelia* represented a distinct branch well separated from the other spirochaetes but closest to the main *Spirochaeta* cluster (Fig. 7.2).

Other Organisms. Host-associated spirochaetes and spirochaete-like organisms have been reported from over 50 species of freshwater and marine molluscs, other aquatic creatures such as starfish, over 40 species of termites, and the wood-eating cockroach. They are of various sizes and some have distinctive morphological features that are visible by electron microscopy, but since none have been grown in pure culture few taxonomic data are available. One monospecific genus, *Cristispira*, is recognized; its members have large, loosely and irregularly coiled cells whose bundles of over 100 flagella may distend the outer sheath to form a ridge (the crista), and they are principally found in the crystalline style and gut of healthy molluscs, where they are assumed to be commensals.

Leptospiraceae

Leptospira, until recently the only genus in this family, contains aerobic spirochaetes which have tightly coiled, slender cells with two periplasmic flagella. Unlike those of other spirochaetes, these flagella rarely overlap in the middle of the cell and their insertion organelles are of the type found in Gram-negative organisms. Other features that distinguish the family were mentioned above.

For many years the genus was split into two species which represented biological groups: *L. interrogans* contained the parasitic strains which tend to localize in hosts' kidneys and cause diseases of varying severity in a wide range of animals, and *L. biflexa* contained the free-living strains, which grow at lower temperatures than *L. interrogans*, and are found in water and moist soil. Traditional approaches to classification were of limited value: there were some biochemical, physiological and other kinds of characters to distinguish between the species but few to differentiate strains within them, and so the basic taxon was the serovar. However, DNA relatedness, restriction endonuclease analysis, and GLC of FAMEs offer more accurate and sensitive approaches to classification and identification, and digest patterns in gels or fatty acid profiles may be subjected to computer taxonomy. Until recently there were about 180 serovars in *L. interrogans* and these were arbitrarily

Table 7.2 Subdivisions of the leptospires*

Species	Mol% G+C†	Pathogenicity	Lipase activity	Growth at 13°C	Serum needed‡	Resistant to: 8-AZA§	Resistant to: 2,6-DAP¶
L. interrogans	34.9	+	+	–	+	–	(–)
L. borgpeterseni	39.8	+	–	–	+	–	(+)
L. inadai	42.6	+	+	–	+	+	–
L. noguchi	36.5	+	(+)	–	+	–	–
L. santarosai	40.7	+	–	–	+	–	(+)
L. weilii	40.5	+	–	–	+	–	–
L. kirschneri		+		–	+	–	(–)
L. biflexa	37.0	–	+	+	+	+/–	+/–
L. meyeri	34.3	–	+	+	+	+	+
L. wolbachii	37.2	–	+	+	+	+	+
L. parva	48.7	–	+	+	+	+	–
Leptonema illini	54.2	–	+	+	–	+	–

* From Yasuda *et al.* (1987) & Ramadass *et al.* (1992).
† Of type or representative strain.
‡ For growth *in vitro*.
§ 8-Azaguanine.
¶ 2,6-Diaminopurine.
+, positive; –, negative; (+), usually positive; (–), usually negative; +/–, variable.

assigned to 19 serogroups (which had no taxonomic standing) on the basis of cross-reacting agglutinogens; *L. biflexa* had been studied less, but it was also serologically heterogeneous and nearly 70 serovars in 38 groups had been reported.

The mol% G+C range of *L. biflexa* and *L. interrogans* was 35–41. A further species, "*L. illini*", differed from the other leptospires in having a mol% G+C of 54, flagella insertion organelles of the type found in *Borrelia*, *Spirochaeta*, *Treponema* and Gram-positive organisms, cytoplasmic tubules as found in treponemes, and in its cultural requirements (Table 7.2) and the new genus *Leptonema* was proposed to accommodate it. The new species *Leptospira parva* was proposed for a non-pathogenic water organism which is biochemically intermediate between the two established species of the genus and has a mol% G+C of 48.

In DNA–DNA hybridization studies *L. biflexa* and *L. interrogans* showed little relatedness and were found to be very heterogeneous. Further DNA homology data confirmed the taxonomic status of *Leptonema illini* and *Leptospira parva* and supported the recognition of five new genomic species of parasitic serovars and two of saprophytic serovars (Table 7.2); another pathogenic species, *L. kirschneri* was proposed following further DNA homology work. Not surprisingly, the new species do not equate with the serogroups — genetically related organisms often being antigenically different — but the FAME profile groups so far recognized show general agreement with the proposed species. A larger study of DNA relatedness among *L. biflexa* strains recognized six rather than two groups but 60% of the strains were left ungrouped. Nevertheless, further work of this kind will allow a taxonomic structure to be established for the family and should provide a useful framework for investigating phenotypic relationships and grouping the serovars.

Studies of rRNA gene restriction patterns reveal that genomic species always have different patterns and that many serovars give specific ones. Although some serovars could not be separated in this way, the method is clearly of value for molecular typing. In 16S rRNA sequence comparisons strains of *L. biflexa* and *L. interrogans* show a closer evolutionary relationship to each other than they do to *Leptonema illini*, and the three represent a branch quite distinct from the other spirochaetes (Fig. 7.2).

Important species

Spirochaetes are major parasites and pathogens of humans and other animals. Epidemiologies are diverse, but in many cases the organisms gain entry through skin or mucous membranes, then appear in the blood and become widely disseminated, causing multistage diseases which may have periods of latent infection, and show tropisms for the skin, heart and central nervous system.

Humans are the natural hosts of *Treponema pallidum* subsp. *pallidum*, the cause of venereal and congenital syphilis, *T. pallidum* subsp. *endemicum*, the cause of endemic syphilis, *T. pallidum* subsp. *pertenue* and *T. carateum*, the respective agents of the tropical skin diseases yaws and pinta, and *Borrelia recurrentis* and *B. duttonii*, which are respectively responsible for the epidemic louse-borne, and one of the endemic tick-borne, forms of relapsing fever. *Borrelia burgdorferi* and *B. garinii*, the agents of Lyme disease, and the various species responsible for the other endemic forms of relapsing fever are primarily tick-borne parasites of animals. Other species of importance include *B. thieleri*, which causes cattle, horse and sheep borreliosis, *B. anserina*, which causes a fatal disease of poultry, and *B. coriaceae*, which is associated with bovine epizootic abortion. *Serpulina hyodysenteriae* is responsible for swine dysentery.

Animals, especially rodents, are also the natural hosts for leptospires. The organisms are transmitted by contaminated urine, and infections in humans range from mild fever to septicaemia with severe liver and kidney involvement.

Identification

The cultivable species of *Spirochaeta* and *Treponema* may be identified using morphological, physiological and biochemical tests and chemotaxonomic methods such as GLC–MS of fatty acids and carbohydrates, but the identification of the other spirochaetes is less straightforward. As may be imagined, much effort has been expended on developing diagnostic tests for the pathogenic species, especially *T. pallidum* subsp. *pallidum*.

In the diagnosis of treponemal infection, dark-ground microscopy is valuable, and many different serological tests of varying specificity, sensitivity, and simplicity have been developed; rapid microhaemagglutination tests are available in kits. However, because the species are morphologically identical and no species-specific antigen has been discovered, differentiation depends on clinical manifestations.

Borrelia burgdorferi cultures can be identified serologically, but many other *Borrelia* species are antigenically unstable, hence the relapsing natures of the fevers, and this makes their serological identification very difficult. For diagnosis, microscopy of blood is useful during febrile periods, and indirect immunofluorescence is of value, but identification to species level is usually only presumptive and depends upon geographical distribution and the specific relationships that occur between some species and their vectors. Genetic probes derived from flagellin gene sequences show promise for identification of *Borrelia* species and taxonomic work within and between them.

As already indicated, the distinction of the two established species of *Leptospira*, *L. interrogans* and *L. biflexa*, and of *L. parva* is quite simple, but this is not the case with the recently proposed species. Differentiation of the serovars is a specialized task, using microscopic agglutination and agglutinin adsorption tests with serovar-specific antisera raised in rabbits, and is largely restricted to reference laboratories, but the genomic methods outlined above appear to have the potential to supplant serological approaches.

8 Helical and curved bacteria

The elegance of form of the organisms exerted a fascination not stimulated by the usual bacteria to which one is so accustomed.

M.A. Williams & S.C. Rittenberg, 1957, *A Taxonomic Study of the Genus Spirillum* Ehrenberg

Introduction

This group contains 10 genera whose members have similar cellular morphologies and share some other characteristics. Compared with the spirochaetes, however, it is not well defined phenotypically or phylogenetically and some parts are best regarded as loose assemblages which require comprehensive genotypic study before satisfactory arrangements can be made.

Members of the group typically have inflexible, Gram-negative, helically curved cells whose numbers of turns range from less than one to many. Emphasis on this morphological character has been tempered by the occurrence of straight rod variants in some species and the existence of organisms which share several characters with *Aquaspirillum* species but normally occur as straight rods. Other characters shared by many members of the group include polar flagella borne at one or both ends of the cell, polar membranes which tend to underlie the areas of flagella insertion, the formation of coccoid bodies in older cultures, aerobic or microaerophilic growth, and inability to attack carbohydrates. The helical bacteria range from saprophytes in soil, fresh water and marine environments through nitrogen fixers growing in association with plant roots, and predators on other bacteria and on algae, to parasites of the mouth, gastrointestinal tract and reproductive organs of humans and other animals.

Classification

The genus *Spirillum* was proposed by Ehrenberg in 1832 to accommodate a rigid spiral organism which he called *S. volutans*. Over the ensuing 140 years about 20 further species were added to the genus which was then divided into the three genera *Spirillum*, *Aquaspirillum* and *Oceanospirillum* on the basis of G+C contents and physiological properties. The eighth edition of *Bergey's Manual*, published in 1974, included *Campylobacter* (a genus which subsequently attracted much interest) in a family of convenience, the *Spirillaceae*. The 1984 edition of *Bergey's Manual* included the well established genus *Bdellovibrio* and the more recently described genera *Azospirillum* and *Vampirovibrio* in the group, but the family name was dropped. Although several apparently important features were shared by most of its members, exceptions made impossible the satisfactory definition of the group as a family.

The helical bacteria are usually inert in traditional biochemical tests so that phenotypic characterization is of limited value in their classification and genomic analyses are clearly required. The G+C range of 30 to 70 mol% represented heterogeneity within the genera as well as low relatedness between them, and nucleic acid hybridization and sequencing studies have already revealed considerable taxonomic complexity in most cases.

It must be appreciated that, despite the emphasis on cellular morphology, genera containing helically curved organisms are to be found in several other groups; examples include anaerobes such as *Desulfovibrio*, facultative anaerobes such as *Vibrio*, and the phototrophs *Rhodospirillum* and *Thiospirillum*. Furthermore, rods curved in one plane are common in many genera.

Spirillum

This genus contains the single species *S. volutans*, which is a large, microaerophilic, freshwater organism with a G+C content of 38 mol%. Its DNA base composition is not much higher than that of *Campylobacter fetus*, to which it shows some phenotypic similarity in its polar flagella and polar membrane, oxygen requirements, respiratory metabolism, and inability to catabolize carbohydrates, but the cells of *S. volutans* are larger and multitrichous. The two organisms also have quite different habitats. Both belong to the *Proteobacteria*, but campylobacters represent a distinct group (superfamily VI) while *S. volutans* belongs to the β group (superfamily III) and, according to 16S rRNA sequencing studies, represents a deeply branched line from the base of the *Alcaligenes–Rhodocyclus–Nitrosomonas* subgroup (Fig. 8.1).

"Spirillum minus", which is a cause of rat-bite fever, and *"S. pulli"*, a pathogen of chickens, do not belong to this genus but their true affiliations are unknown.

Aquaspirillum

Like *Spirillum* these are freshwater organisms which will not tolerate even moderately salty conditions. They differ in being much smaller and aerobic, and their wide phenotypic diversity and range of phylogenetic relationships, and a much higher G+C range of 49–66 mol%, showed *Aquaspirillum* to be the least satisfactory genus in the group. As well as typical spirilla with bipolar tufts of flagella, the genus contained a non-helically curved species with one or two flagella at only one pole (*A. delicatum*) and one with straight rods (*A. fasciculus*). Although they are all chemo-organotrophs at least one is a facultative hydrogen autotroph; they grow with low nutrient levels and most are unable to utilize carbohydrates, so amino acids and other organic acids are used as carbon sources. These and other physiological and biochemical 'core' characters were taken, along with the mol% G+C range, to define the genus. Variations in some of them and the possession of rarer properties such as the formation of coccoid bodies and nitrogenase activity were, of course, useful to distinguish

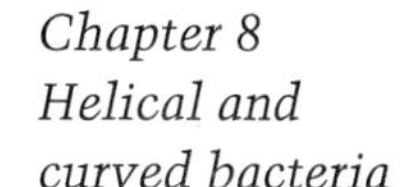

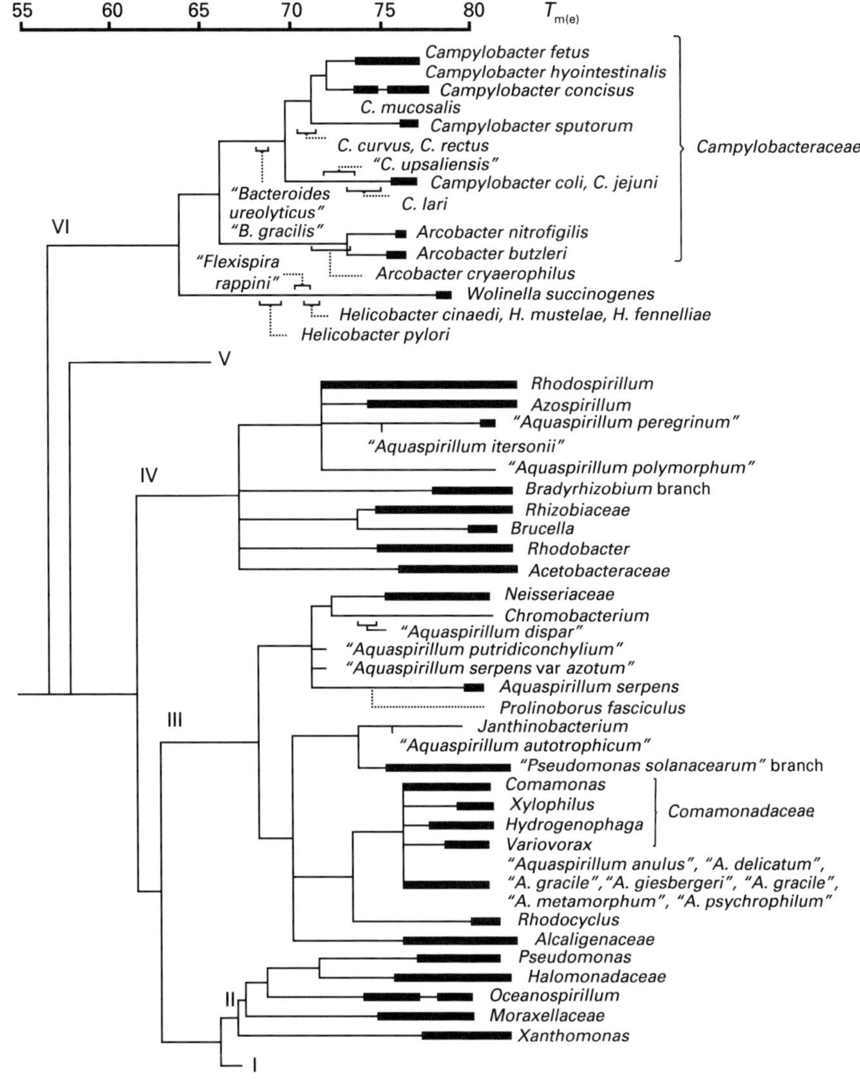

Fig. 8.1 rRNA cistron similarity dendrogram showing the relationships of the helical and curved bacteria. Adapted from Vandamme *et al.* (1991).

between the 18 species, which were recognized mainly on the basis of morphological and nutritional differences and DNA base composition.

A comprehensive DNA–rRNA hybridization study revealed at least a dozen distinct groups of *Aquaspirillum* species which were intermingled with members of the α and β proteobacterial subclasses (superfamilies IV and III) (Fig. 8.1). It was proposed that the genus be restricted to the type species, *A. serpens*, and its biovars. Surprisingly, its closest relative is the rod shaped species *A. fasciculus*, which was placed in the new genus *Prolinoborus*. These, along with "*A. dispar*" and "*A. putridiconchylium*", are members of the *Neisseriaceae–Chromobacterium* branch of superfamily III whereas the other aquaspirilla in the superfamily ("*A. aquaticum*" – now *Comamonas aquaticum*, "*A. anulus*", "*A. delicatum*", "*A. giesburgeri*", "*A. gracile*", "*A. metamorphum*", "*A. psychrophilum*" and "*A. sinuosum*") represent new or existing genera in the *Comamonadaceae*. The species "*A. itersonii*", "*A. peregrinum*" and "*A. polymorphum*" belong to superfamily IV, where their closest relatives are *Azospirillum* and *Rhodospirillum*, as do the magnetotactic magnetogens of *Magnetospirillum*. Other nomenclatural changes await the study of larger numbers of strains and more phenotypic characters. Further nucleic acid studies, especially DNA–DNA hybridizations, will reveal the relative significances of the morphological and physiological core characters, allowing phenotypic differentiation of new taxa.

Azospirillum

Azospirillum species are microaerophilic nitrogen-fixing organisms that occur free in the soil or in association with grasses and tuberous plants. Unlike the other spirilla they may produce numerous lateral flagella in addition to single polar flagella. Their mol% G+C range is 64–71, and DNA–rRNA and DNA–DNA hybridizations, and restriction endonuclease pattern analyses show them to be closely related but distinct. In addition, they selectively infect the roots of different ranges of host plants, and are easily distinguished by phenotypic tests.

Their closest known relatives in the α subclass of the *Proteobacteria* are *Rhodospirillum rubrum* and its relatives, and misnamed *Aquaspirillum* species (Fig. 8.1). Although the five azospirilla share some phenotypic similarities (such as vibrioid morphology, growth requirements, microaerophilic nitrogen fixation and limited attack on carbohydrates) with certain of the aquaspirilla, the recognition of two genera is supported by the typical characters of the latter group, as well as by G+C ranges and nucleic acid homologies. With regard to *R. rubrum*, some azospirilla also produce pink/red pigments, and members of both taxa form intracellular poly-β-hydroxybutyrate, and have the same unique activating factor for their nitrogenase Fe proteins, but their cellular morphologies and G+C ranges differ. *Azospirillum* occupies the same proteobacterial group as *Beijerinckia*, *Rhizobium* and *Xanthobacter*, but it is not closely related to these or to any of the other genera of nitrogen fixers investigated.

Oceanospirillum

This genus was established to contain five species of aerobic, marine spirilla which required seawater for growth and spanned a range of 42–48 mol% G+C; three further species extended the G+C limit to 51 mol%. Like *Aquaspirillum*, this genus was defined on the basis of patterns of phenotypic characters: cell shape, flagellation, formation of coccoid bodies, respiratory metabolism, inability to catabolize carbohydrates, simple heterotrophic nutrition and optimum temperatures of around 30°C. It became clear that the genus was somewhat heterogeneous, though not to as great an extent as *Aquaspirillum*. Oligonucleotide cataloguing showed that the species *O. maris* and *O. minutulum* were not closely related although they both belonged to the same subgroup of the γ group of the *Proteobacteria*.

The upper limit of the G+C range was extended to 57 mol% and the definition of the genus was drastically changed by the proposed inclusion of two former *Alteromonas* species so that many useful phenotypic characters were dropped. This proposal was based upon immunological studies of Fe-containing superoxide dismutase and glutamine synthetase; however, the same species had been assigned to the new genus *Marinomonas* by the Gent group on the basis of DNA–rRNA hybridizations. This unsatisfactory state of affairs was rectified by a comprehensive, polyphasic study by the Gent group, in which relationships at the generic level and above were revealed by DNA–rRNA hybridizations, and relationships between and within species were investigated by DNA–DNA hybridization and gel electrophoresis of whole-organism proteins.

These mainly genotypic studies indicated the significances of several phenotypic characters, and essentially led to a return to the original definition of the genus, but with a G+C range of only 45–50 mol%, and to the exclusion of two species and the classification of two others as subspecies. The species excluded were the small-celled organisms "*O. minutulum*" (42–44 mol% G+C) and "*O. pusillum*" (51 mol% G+C), which were the only nitrate-reducing members of the genus. The affiliation of the former within the γ subclass of the *Proteobacteria* is unknown; the latter, which differs from the other oceanospirilla in having cells with anticlockwise helices and single polar flagella rather than tufts, belongs to the α subclass and is related to misnamed *Aquaspirillum* species, *Azospirillum* and *R. rubrum*.

Campylobacter

The genus *Campylobacter* (meaning curved rod) was created to accommodate several vibrioid agents of bovine and ovine reproductive diseases which are unlike *Vibrio* species in being microaerophilic organisms which do not attack carbohydrates and have a much lower mol% G+C range. Interest in the genus has greatly increased since the mid-1970s when it was demonstrated that *Campylobacter* species could be isolated from the faeces of many patients suffering from enteritis by cultivation on a selective medium at 42°C in a microaerobic atmosphere. Alongside a dramatic rise in reported clinical isolations in most parts of the world (Fig. 8.2), the number of species has increased. The 1974 edition of *Bergey's Manual* recognized three species and five subspecies, in 1984 there were five species and five subspecies, and by 1993 the number of valid and proposed species had risen to 12 with four subspecies, and the establishment of two new genera for 12 other species.

Campylobacter species not only fail to attack carbohydrates but are inert in most other traditional biochemical tests, and this has made their separation difficult. Several species and subspecies are separated on the basis of very few tests, sometimes only one; *C. jejuni* and *C. coli*, for example, are separated by hippurate hydrolysis, the other few tests being unreliable, and the subspecies of *C. fetus* are distinguished only by their ability to grow in the presence of 1% glycine. None the less, studies of DNA relatedness have

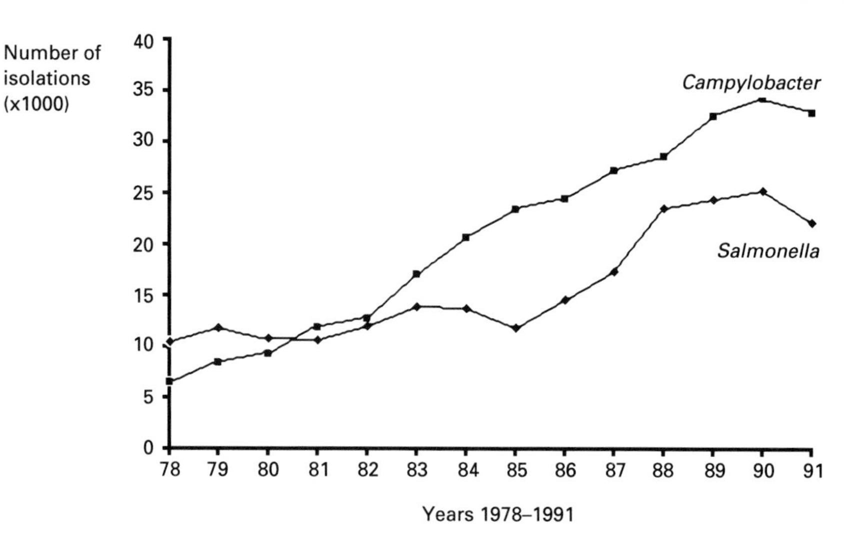

Fig. 8.2 Laboratory isolations of *Campylobacter* and *Salmonella* in England and Wales, reported to the Communicable Disease Surveillance Centre from 1978 to 1991.

largely upheld the recognition of these taxa and have enabled the selection of differential phenotypic tests. The distinction between *C. fetus* subsp. *fetus* and *C. fetus* subsp. *venerealis*, both of which cause bovine and ovine abortion, was not supported on genotypic grounds but is of practical epidemiological value as they are transmitted by different routes.

In addition to DNA relatedness, numerical analysis of DNA restriction endonuclease digest patterns has achieved separation of several species and their biotypes. Also, the results of chemotaxonomic studies and ultrastructural observations have been most revealing. Profiles of whole-organism proteins, fatty acids and isoprenoid quinones are generally characteristic for the well-established species, and those for several of the more recently described species indicate that they have been incorrectly assigned.

These conclusions have been supported by the results of phylogenetic approaches using oligonucleotide cataloguing and partial sequencing of 16S rRNA, which revealed three homology groups. One (with a range of 29–38 mol% G+C) contained the type species, *C. fetus*, and all the other well-established species along with several newer species and four putative anaerobes (which had been assigned to *Bacteroides* and *Wolinella*). The other two groups were considered to represent different genera (Table 8.1). It is clear that the phenotypically defined working groups of thermophilic campylobacters, and catalase positive and catalase negative or weak (CNW) types, do not correlate with the phylogenetic groupings.

Over the years various unidentified spirochaete- and spirillum-like bacteria have been observed in the mouths and guts of humans and other animals. One such organism was found in close association with gastritis and named *C. pyloridis* (subsequently changed to the grammatically correct *C. pylori*), but its chemotaxonomic profiles are quite different to those of true *Campylobacter* species. Whole-organism protein patterns were similar for *C. pylori* strains isolated in Australia and England, and other studies found distinctive fatty acid and menaquinone profiles.

Table 8.1 Characteristics of some *Campylobacter* species and their relatives

Species*	16S rRNA homology group & subgroup†	23S rRNA super-family VI cluster‡	Mena-quinones§	Growth at 42°C	Catalase	Sheathed multiple polar flagella
C. fetus						
subsp. *fetus*	I 1	I	*MK-6	v	+	–
subsp. *venerealis*	I 1	I	*MK-6	–	+	–
C. hyointestinalis	I 1	I	*MK-6	v	+	–
C. concisus	I 1	I	*MK-6	–	–	–
C. mucosalis	I 1	I	*MK-6	+	–	–
C. sputorum	I 1	I	*MK-6	+	–	–
C. coli	I 2	I	*MK-6	+	+	–
C. jejuni	I 2	I	*MK-6	+	+	–
C. lari	I 2	I	*MK-6	+	+	–
"*C. upsaliensis*"	I 2	I	*MK-6	+	–/w	–
H. cinaedi	II	III	Un-MK-6	–	+	+
H. fennelliae	II	III	Un-MK-6	–	+	+
H. pylori	II	III	Un-MK-6	–	+	+
H. mustelae	II	III	*MK-6	+	+	+¶
W. succinogenes	II	III	*MK-6	w	–	–
A. cryaerophilus	III	II	Un-MK-6	–	+	–
A. nitrofigilis	III	II	Un-MK-6	–	+	–

* *Campylobacter, Helicobacter, Wolinella* and *Arcobacter.*
† From Thompson *et al.* (1988).
‡ From Vandamme *et al.* (1991).
§ *MK-6 is 2,[5 or 8]-dimethyl-3-farnesyl-farnesyl-1,4-naphthoquinone and Un-MK-6 is an unidentified quinone; all species possess MK-6 which is 2-methyl-3-farnesyl-farnesyl-1,4-naphthoquinone.
¶ Also has lateral flagella.
v Variable within species; w, weak.

The major fatty acids of *C. pylori* are tetradecanoic (14:0) and *cis*-9,10-methyleneoctadecanoic acids (19:0Δ) with a little hexadecanoic acid (16:0), whereas true *Campylobacter* species have hexadecanoic, octadecenoic (18:1) and hexadecenoic (16:1) acids. *Campylobacter pylori*, like several other recently described species, lacks the methylated menaquinone *MK-6 that is characteristic of the genus but contains an unidentified quinone (Un-MK-6) (Table 8.1).

On the basis of 16S rRNA sequencing it was suggested that *C. pylori* should be placed in *Wolinella*, but the type species and supposed close relative of *C. pylori*, *W. succinogenes*, shows ultrastructural dissimilarities to it, has fatty acid and menaquinone profiles closer to those of true campylobacters (Table 8.1) and has the much higher range of 44–49 mol% G+C compared with 36–38 mol% for *C. pylori*. As a result of a polyphasic approach to taxonomy, with contributions from conventional cultural, physiological and biochemical tests, miniaturized enzyme tests, ultrastructural observations, fatty acid and menaquinone profiles, DNA composition and DNA–DNA hybridization, *C. pylori* was placed in the new genus *Helicobacter* along with *C. mustelae*, which is a related organism from the gastric mucosa of ferrets. Although *Helicobacter* was only established in 1989, it comprised eight species from the gastrointestinal tracts of various animals by 1993.

A major phylogenetic study based upon 23S DNA–rRNA hybridizations also revealed three groups (Fig. 8.1 & Table 8.1) and it was concluded that *W. curva* and *W. recta* should be transferred to *Campylobacter*, *C. cinaedi* and *C. fenelliae* to *Helicobacter*, and that a new genus, *Arcobacter*, be proposed for *C. cryaerophila* (becoming *A. cryaerophilus*) and *C. nitrofigilis*. The new species "*C. butzleri*", proposed at about the same time, belongs in *Arcobacter*.

Campylobacter now appears to be a well defined genus and, together with *Arcobacter*, forms the new family *Campylobacteraceae*; along with its relatives *Helicobacter* and *Wolinella*, and the sulphur-oxidizer *Thiovulum*, it represents a distinct branch in the proteobacterial phylogenetic tree (superfamily VI). However, other helical, gut organisms await isolation and characterization ("*Gastrospirillum hominis*" for example) and will no doubt pose further taxonomic problems so that it may be some time before other genera in this group can be satisfactorily defined.

Bdellovibrio

The remarkable intracytoplasmic predatory nature of the members of this genus is its main defining feature, the name *Bdellovibrio* meaning leech-like vibrio. The small, aerobic, vibrioid cells are highly motile by single sheathed polar flagella and on striking Gram-negative prey cells they attach and then penetrate into the periplasmic space. The host soon dies as the predator grows into a helical cell which then fragments into more motile vibrioid forms. Prey-independent and facultative predator strains have been derived *in vitro* and they also produce the fragmenting helical forms.

Although the special behaviour of the bdellovibrios has stimulated much study, conventional approaches to their classification have been inhibited by the necessity of cultivation in living hosts. Even when host-independent strains are available metabolic limitations restrict the value of traditional characterization tests. Of the physiological and biochemical properties that have been considered for their taxonomic values, only protease activity, catalase, nitrate reduction and sensitivity to vibriostatic agent O/129 are known to be reliable. There is at least one common antigen, but the three species are antigenically distinct and are also divisible by bacteriophage susceptibility; the type species *B. bacteriovorus* is divisible into nine serovars and four phage groups, which may support subdivision of the species.

The recognition of more than one species in the genus has relied upon analysis of G+C content, genome size, DNA–DNA hybridization and host range. The last of these has not been investigated thoroughly, but strains of *B. starrii* are noted for their failure to attack enterobacteria. Cytochrome spectra have also been studied and appear to be of some value. The limited DNA homology data show the three established species to be distinct and indicate genetic diversity within the genus, but no more than is found in many other genera. The G+C range for these freshwater species is 42–51 mol% but some incompletely characterized marine isolates push it down to 33 mol%.

Oligonucleotide cataloguing of 16S rRNA indicated that the freshwater species belong to the δ group of the *Proteobacteria* where they form a group distinct from the other member groups, the sulphur and sulphate reducers such as *Desulfovibrio* and the myxobacteria. *Bdellovibrio bacteriovorus* appears to be a rapidly evolving organism, however, and so its position is uncertain.

Other predatory bacteria, whose relationships to *Bdellovibrio* are unknown, include a marine ectoparasite of bacteria and ecto- (*Vampirovibrio*) and endoparasites of algae.

Important species

Of the organisms considered in this chapter *Campylobacter* species stand out because of their veterinary and medical importance. They are common gut commensals of animals and several are important pathogens; *C. fetus* subsp. *fetus* causes bovine and ovine abortion, *C. fetus* subsp. *venerealis* causes bovine abortion and infertility, *C. mucosalis* and *C. hyointesinalis* are associated with intestinal disease in pigs, and other species cause abortion, enteritis and mastitis in domestic animals.

Since routine screening and reporting began in some countries in the late 1970s, and as isolation methods have improved, campylobacter enteritis, mainly caused by *C. jejuni*, has emerged as a pre-eminent worldwide zoonosis, and in Britain reports have outstripped those of salmonellosis (Fig. 8.2). Infection is mainly from consumption of contaminated food, milk and water, with poultry as a major source in developed countries. Various species, especially *C. fetus*, may also cause systemic infections. *Helicobacter pylori* infection is associated with gastritis, duodenitis and peptic ulcer disease, and its role in these conditions is the subject of intensive study.

Identification

As indicated earlier, there is a paucity of conventional tests suitable for separating the species of the various groups. After morphological observations and ecological considerations, identification relies on a few physiological and biochemical tests such as catalase and nitrate reduction, along with growth conditions and temperatures, and resistance to inhibitory agents.

Much attention has been focused on the identification of the pathogenic campylobacters and the *in vivo* detection of *H. pylori*. Like many other gut bacteria, campylobacters show great antigenic diversity; as no satisfactory species-specific group antigens have been identified, serotyping is principally of epidemiological value. Other schemes, useful for identification as well as for epidemiological typing, include: biotyping using tests for hippurate and DNA hydrolysis, H_2S production, and resistance to various antibiotics, dyes and other chemicals; numerical analysis of PAGE protein profiles; plasmid analysis by agarose-gel electrophoresis; and DNA restriction endonuclease analysis. More recently, methods based upon nucleic acid probes for the DNA that encodes rRNA have been developed for species identification.

9 Gram-negative aerobic bacteria

Nomenclature should reflect genomic phylogenetic relationships to the greatest extent possible and . . . all preconceived notions ought to be reexamined within this context.

International Committee for Systematic Bacteriology, 1990, *Ad Hoc Committee Report*

Introduction

The title of this chapter does not refer to a natural taxonomic group. *Bergey's Manual of Systematic Bacteriology* included 35 eubacterial genera in Section 7, the 'Gram-Negative Aerobic Rods and Cocci', but this interim arrangement was mainly for practical purposes, pending comprehensive phylogenetic studies; while 19 of the genera were assigned to seven families, the affiliations of the other 16 genera remained unknown. However, although aerobic metabolism appears to have arisen several times in the course of bacterial evolution and the genera to be considered are found to be intermixed with anaerobes, all those that have been characterized belong to the α, β and γ subclasses of the *Proteobacteria*. None the less, some families contain genera from several proteobacterial subdivisions and species of *Pseudomonas* were similarly broadly spread prior to recent nomenclatural revisions. Problems such as these make the integration of the existing phenotypic classifications and the emerging phylogenies a very difficult and sometimes impossible task. The new phenotypic arrangements need to be based upon high-quality data for characters whose significances are indicated by phylogenetic analyses. This is well illustrated by consideration of some of the groups discussed below.

"Pseudomonadaceae"

This used to be a very large family, 25 genera being proposed for inclusion at one time, but increases in knowledge of the natural relationships between bacteria allowed a reduction to four genera with a mol% G+C range of 58–71: *Frateuria, Pseudomonas, Xanthomonas* and *Zoogloea*. It is defined as a family of strictly aerobic, straight or curved chemo-organotrophic rods which are motile by polar flagella and have respiratory metabolisms. The type genus *Pseudomonas* was defined much as for the family, with a few characters such as inability to grow in acid conditions, the production of xanthomonadin pigments, and production of floc-forming gelatinous matrices serving to separate *Frateuria, Xanthomonas* and *Zoogloea* respectively.

Despite this circumscription the family remained unsatisfactory. *Zoogloea* was only tentatively included and the taxonomy of this genus was somewhat confused until a new neotype strain was designated (the original description having been taken from a mixed culture), and the growth habit is a rather overemphasized character. Furthermore, this organism doesn't belong to any of the six RNA superfamilies recognized to date. Flagellation was also overemphasized and species with lateral, sheathed, or no flagella were all accepted in *Pseudomonas*, although several of these species show stronger evolutionary relationships with peritrichously flagellate organisms than they do with other species of the genus. *Frateuria* and *Xanthomonas* are certainly related, but they lie at some distance from the true pseudomonads.

Pseudomonas

Typical pseudomonads are easily recognized: aerobic, oxidase-positive rods which are widely distributed in the environment and which have single polar flagella and the ability to utilize wide ranges of simple organic compounds. For many years the taxonomy of the genus was, however, very

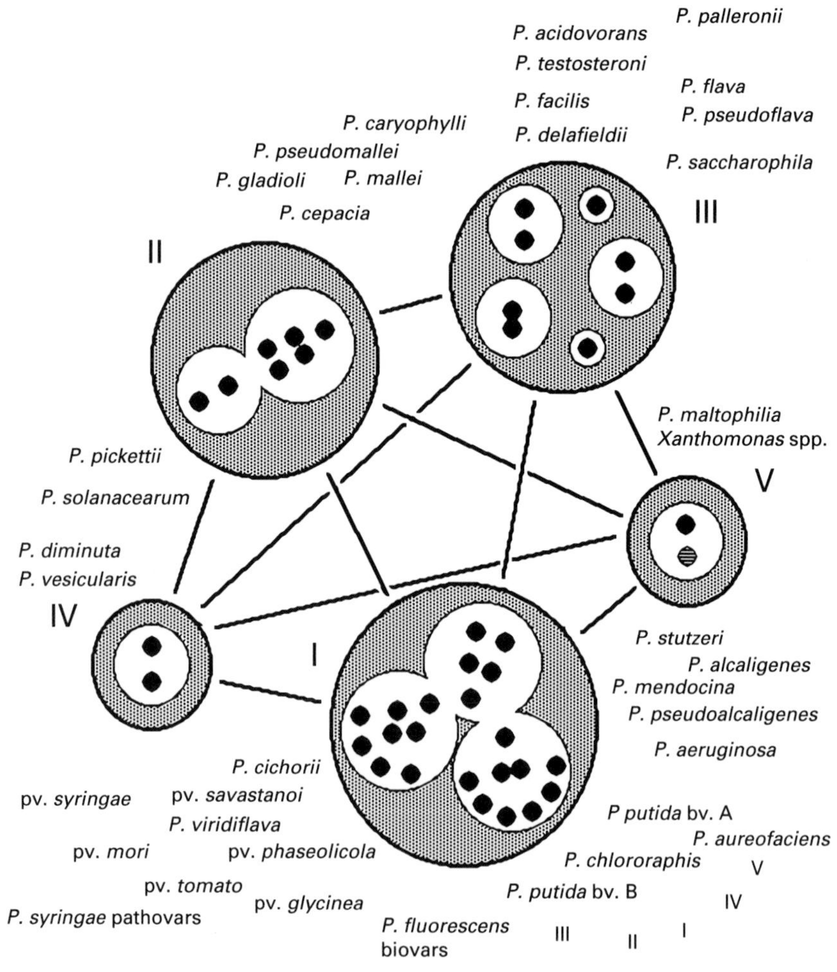

Fig. 9.1 *Pseudomonas* species, biovars and pathovars arranged according to rRNA and DNA relatedness. The large shaded circles indicate rRNA relatedness groups, and the white circles within them indicate DNA relatedness groups. Adapted from Palleroni (1984) *Bergey's Manual of Systematic Bacteriology*, Volume 1, p. 161.

complex; there were nearly 100 species and numerous plant pathovars which were subsequently found to exhibit a range of phylogenetic origins.

Many species had been extensively characterized by conventional phenotypic tests, while others awaited such thorough study. Initially, taxa were recognized by visual examination of the data, and numerical methods were only applied to limited numbers of strains in certain parts of the genus. However, the groupings arrived at correlated very well with those indicated by the early DNA–DNA and DNA–rRNA hybridization studies of Palleroni's group at Berkeley (Fig. 9.1 & Table 9.1) and by Sneath's group's subsequent numerical taxonomy based upon published data (Fig. 9.2). The genus was seen to comprise three main rRNA relatedness groups, each containing several DNA relatedness subgroups, with two small, peripheral rRNA groups.

The five pseudomonad groups described below were based on studies of a minority of the species, and although splitting *Pseudomonas* into several

Table 9.1 Correlations of nucleic acid relatedness groups, proteobacterial divisions and phenotypic properties in the pseudomonads

RNA and DNA group	Proteo-bacterial subclass/ RNA super-family	Pigments, ecologies and groups	Phenotypic properties				
			Growth factors needed	Nitrate used as N source	Oxidase test	PHB*	Growth at 40°C
Ia	γ/II	Non-fluorescent saprophytes	–	+	+	–	+
Ib	γ/II	Fluorescent saprophytes, opportunists and phytopathogens (true *Pseudomonas*)	–	+	+/–†	–	– (+)‡
IIa,b	β/III	Animal and plant pathogens (solanacearum group)	–	+	+	+	+
IIIa, b and c	β/III	Saprophytes (*Comamonadaceae*)	–	+	+	+	– (+)
IV	α/IV	Saprophytes ("*P. diminuta*")	+	–	+/w§	+	–
V	γ/II	Xanthomonadins, phytopathogens (*Xanthomonas*)	+	– (+)	–/w	–	–

* Poly-β-hydroxybutyrate accumulated as a carbon reserve material.
† Variable for group.
‡ Positives unusual.
§ Weak reaction.

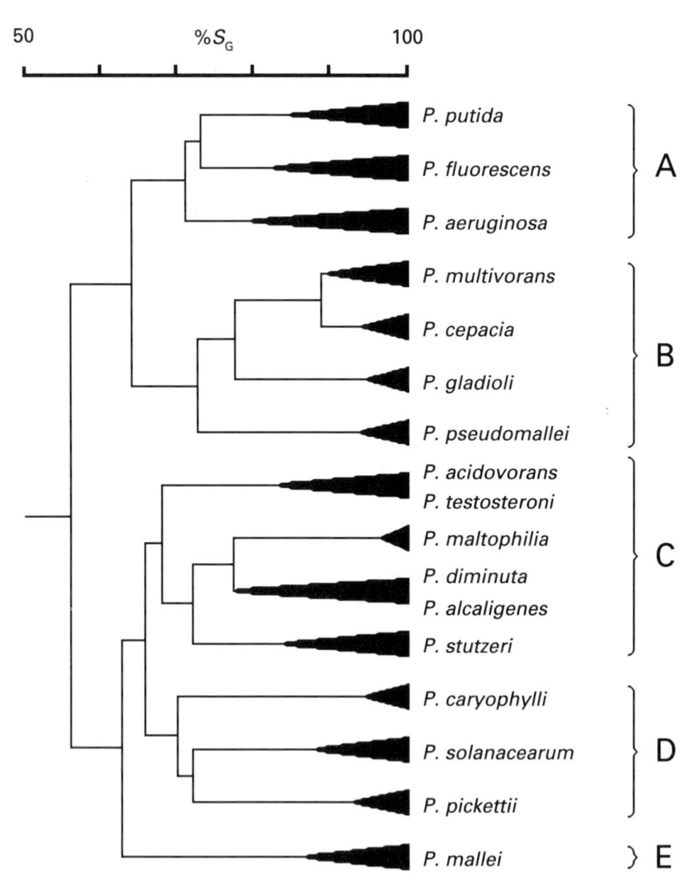

Fig. 9.2 Simplified average linkage phenogram for some *Pseudomonas* species, derived from an analysis of published substrate utilization data by Gower's coefficient. Adapted from Sneath *et al.* (1981).

genera to correspond with the nucleic acid hybridization groups and subgroups is desirable, it is a slow process requiring the characterization of many further species. However, several large polyphasic studies have already resulted in the proposal of some of these groups as separate genera. Certainly, DNA–rRNA hybridization studies and 16S rRNA sequence data fully support division: representatives of group I, the fluorescent pseudomonads, and of group V, the xanthomonads, belong to the γ subclass of the *Proteobacteria* (superfamily II), those of the two closely related groups II and III to different subgroups of the β subclass (superfamily III), and "*P. diminuta*", representing group IV, to the α subclass (superfamily IV) (Table 9.1 & Fig. 9.3).

The Fluorescent or True Pseudomonads. The largest group may be called the fluorescent pseudomonads as many of its members produce fluorescent pigments. It represents the genus *Pseudomonas sensu stricto* as it contains the type species *P. aeruginosa* which, along with several medically important, non-pigmented, denitrifying species such as *P. alcaligenes* and *P. stutzeri*, forms one DNA relatedness group; this group is further divisible

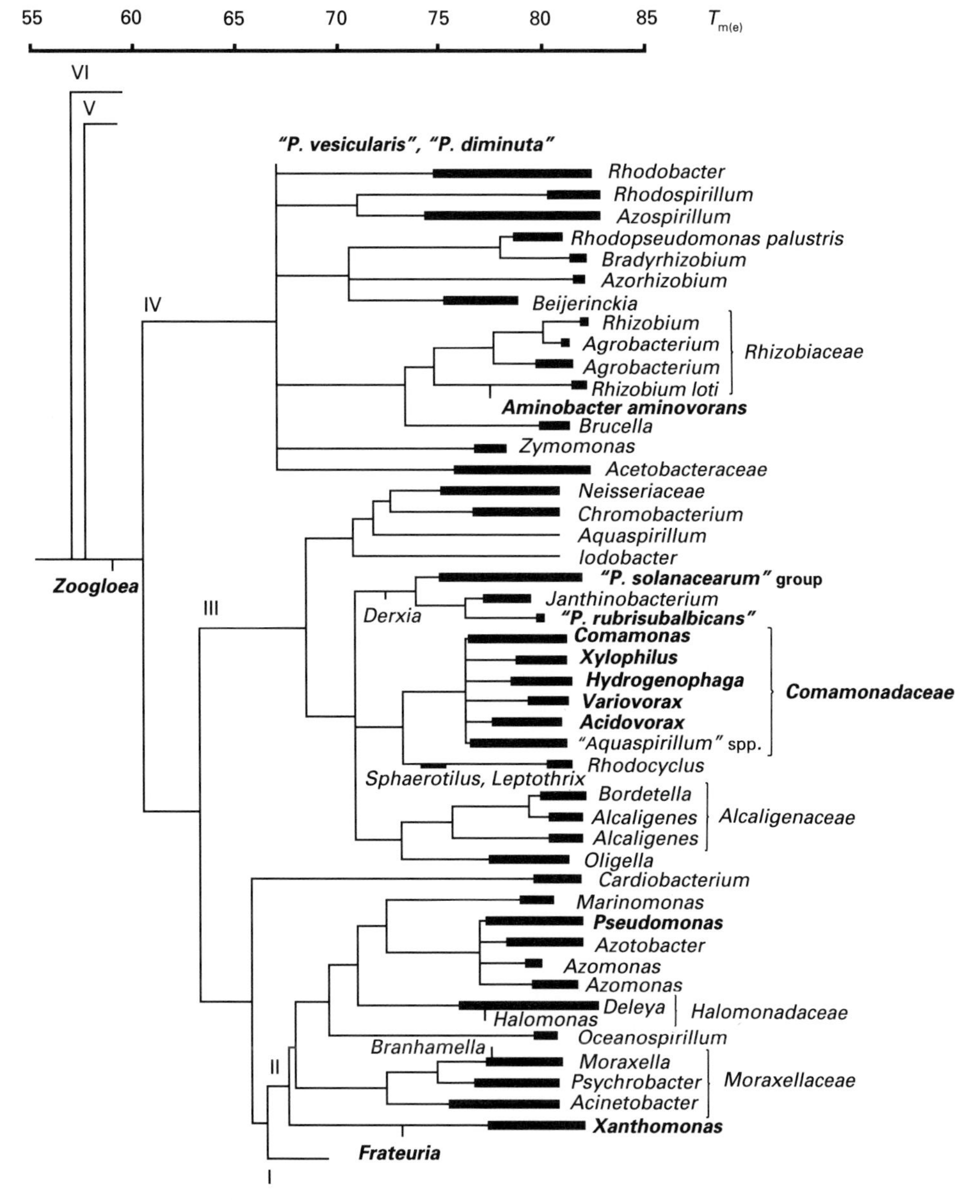

Fig. 9.3 Locations of *Pseudomonas* and other taxa previously assigned to the family *Pseudomonadaceae*, all indicated in bold print, in a rRNA cistron similarity tree.

by numerical taxonomy using conventional tests. A second group contains *P. syringae* and its many pathovars, *P. cichorii* and *P. viridiflava* which are plant pathogens, and the third group mainly comprises saprophytic organisms such as *P. fluorescens* and *P. putida*. The plant pathogens tend to have lower growth rates and temperatures, and lesser nutritional versatilities than the saprophytes.

The neatness of the classification so presented is illusory; the saprophytic pseudomonads are taxonomically very complex and although many environmental isolates can be categorized with some confidence, many others cannot and a considerable amount of further characterization work is needed to define the boundaries of some of the taxa, if such boundaries exist. *Pseudomonas putida* and *P. fluorescens*, for example, are heterogeneous species and have been divided into two and five biovars respectively; a polyphasic study of the latter species by numerical taxonomy of conventional phenotypic characters, DNA–DNA hybridization, and quantitative microcomplement fixation using antibodies to the protein azurin, confirmed the arrangement, with the results of the different approaches showing impressively high congruence.

Pseudomonas aeruginosa and its non-fluorescent relatives are the only valid species in this group that can grow at 41°C, but fluorescent strains distinct from *P. aeruginosa* and closely related to the alcaligenes and stutzeri groups (according to phenotypic tests, whole-organism fatty acids, and DNA relatedness) are also capable of growing at this temperature. This is perhaps not surprising from the evolutionary point of view, but it adds a further complication to the identification of these species.

The taxonomy of the fluorescent plant pathogens is also very complex and for some years was exacerbated by the naming of strains according to limited host specificity testing and the types of lesions produced, and without sufficient other phenotypic characterization. Although plants are of some value as test systems, full host specificity data can never be obtained and they are not usually practicable as identification systems. None the less, although plant testing cannot replace the more conventional approaches to classification, host ranges are of great importance to plant pathologists. Many of the numerous species of fluorescent plant pathogens are therefore regarded as pathovars of *P. syringae* in order to provide a special purpose classification of the kind required by plant pathologists, and it has been found that these pathovars are divisible into six DNA relatedness groups. The species name *P. syringae* does not have priority, but was conserved because it was more familiar than *P. mori* which was published 9 years earlier in 1893.

rRNA Groups II and III — the Acidovorans and Solanacearum Complexes. Being members of the same superfamily, these two groups are closer to each other than they are to group I, from which they are easily separated by conventional tests. Their phenotypic separation is not entirely distinct — important characters such as growth at 41°C and denitrification are described as almost universally positive for group II, for example — but they can be separated in another way. Under strictly standardized conditions, members of group II cleave the aromatic intermediate compound protocatechuate by an *ortho* mechanism whereas group III organisms start with a *meta* cleavage.

A notable feature of group II, which is also known as the solanacearum complex, is the pathogenicity of most, if not all, of its members to plants and animals, ranging from "*P. mallei*", the cause of glanders, which is an equine disease that is transmissible to humans, and its close relative "*P. pseudomallei*", which is a soil organism causing melioidosis in animals and humans, through "*P. cepacia*" which is an opportunistic pathogen of onions and humans, to the important plant pathogens "*P. caryophylli*" and "*P. solanacearum*". The new genus *Burkholderia*, with *B. cepacia* as the type species, has been proposed to accommodate members of this group.

Members of group III, the acidovorans complex, are saprophytes and may be divided into two physiological subgroups: the hydrogen oxidizers (hydrogen pseudomonads) such as "*P. facilis*", "*P. flava*" and "*P. palleronii*" and the species "*P. acidovorans*" and "*P. delafieldii*" which cannot grow chemolithotrophically. This division does not correlate with DNA–rRNA hybridization groups however, and polyphasic studies led to the yellow-pigmented species of the former group ("*P. flava*" and "*P. palleronii*") being placed in the new genus *Hydrogenophaga*, and "*P. delafieldii*" and "*P. facilis*" being transferred to another new genus, *Acidovorax*, along with clinical isolates in the new species *A. temperans*, and plant pathogens of the "*P. avenae*" group. The remaining pseudomonad species in this complex, including "*P. acidovorans*", are accommodated in the emended genus *Comamonas* and the entire complex, encompassing misnamed *Alcaligenes*, *Xanthomonas* (now *Variovorax paradoxus* and *Xylophilus ampelinus* respectively) and *Aquaspirillum* species, forms the new family *Comamonadaceae*.

rRNA Group IV. This group contains the closely related species "*P. diminuta*" and "*P. vesicularis*", which differ from other *Pseudomonas* species in requiring growth factors, in being unable to use nitrate as their source of nitrogen, and in having unique lipid compositions in which sugar-containing lipids are prominent. As they belong to superfamily IV (subclass α) and do not appear to be closely related to any other Gram-negative genera, they probably represent a new genus.

As shown, the groupings indicated by nucleic acid homologies are often supported by other taxonomic approaches such as conventional phenotypic tests, cell wall analyses, fatty acid analysis (especially the 2- and 3-hydroxy and branched 3-hydroxy acids), and immunological characterization of certain proteins or amino acid sequencing. The new genera were proposed

Table 9.2 Cofactors, dehydrogenases, and feedback inhibition in tyrosine biosynthesis in pseudomonad groups

	Activity				
	Prephenate dehydrogenase		Arogenate dehydrogenase		
RNA Group	NAD	NADP	NAD	NADP	Inhibition of activity by tyrosine
I	+	–	+	–	+
II	+	+	+	+	+
III	+	+	+/–	+	–
IV	+	+	+/–	–	–
V	+	–	+	–	+/–

From Byng *et al.* (1980).

following such polyphasic work. Additional and interesting confirmation of the rRNA relatedness groups has been given by investigations of metabolic pathways and their regulatory mechanisms for tyrosine biosynthesis, by determining cofactor specificities and dehydrogenase activities using spectrophotofluorometric measurement of NADH and NADPH formation (Table 9.2). Similar studies on phenylalanine biosynthesis indicated groups which correlate well with those established by nucleic acid hybridizations, as do groups based on control mechanisms for the enzyme 3-deoxy-D-*arabino*-heptulosonate 7-phosphate (DAHP) synthetase.

These groupings were only based on some 35–40 of the well-characterized species and biotypes, leaving nearly twice that number of species awaiting adequate description and allocation. An extensive DNA–rRNA hybridization study has indicated that about one-third of such species belong in RNA group I (the true pseudomonads) and that most of the remainder belong to various proteobacterial RNA branches other than those comprising the *Comamonadaceae*. Many, therefore, are clearly not pseudomonads; for those that are, much further characterization work is required to determine which of them deserve species status and how many genera should be established.

rRNA Group V — Xanthomonas. This genus represents a deep branch of superfamily II (proteobacterial subclass γ) and lies at some distance from *Pseudomonas sensu stricto*. Like the species in group IV, its members require growth factors and do not use nitrate as a nitrogen source. They may be phenotypically distinguished from the other pseudomonads by their production of characteristic yellow pigments (xanthomonadins), which are brominated aryl polyenes. This distinction is confirmed by 16S and 23S rRNA–DNA hybridization studies, which indicate that the genus forms a fifth relatedness group, and by chemotaxonomic analyses; for example, the main polyamine of xanthomonads is spermidine whereas for the *P. fluorescens* group it is putrescine.

All but one of the species are plant pathogens and the type species, *X. campestris*, has nearly 150 pathovars, but cellular protein, fatty acid and DNA restriction endonuclease patterns have shown that several of the pathovars are heterogeneous, and DNA homologies do not always match with host ranges. There is clearly a need for a comprehensive, polyphasic taxonomy, but this will be an enormous task involving pathogenicity testing of each of the groups defined; a numerical taxonomy has, however, supported the recognition of at least eight species. The atypical species is the multitrichous, unpigmented *X. maltophilia*, whose pathogenicity is uncertain. This organism was transferred from *Pseudomonas* on the basis of rRNA and DNA relatedness, and characteristic chemotaxonomic patterns which include the presence of unique, branched hydroxy fatty acids in the cell envelopes, quantitative and qualitative differences in the sugars of the lipopolysaccharides, and unique control mechanisms for DAHP, but it differs from the other xanthomonads in many other respects.

The anomalous species *X. ampelina*, a plant pathogen that differs from the other species in many important characters, is not related to

Xanthomonas and has been placed in a new genus *Xylophilus* on the basis of a numerical taxonomy of chromogenic enzyme tests, PAGE of cellular proteins and 23S rRNA–DNA hybridizations. The fastidious, xylem-limited plant pathogens placed in the new genus *Xylella*, however, are related to the xanthomonads, as shown by 16S rRNA sequencing.

Azotobacteraceae, *Rhizobiaceae* and other nitrogen-fixing bacteria

Biological nitrogen fixation is essential to the maintenance of life on earth and is almost entirely restricted to the prokaryotes, which are responsible for the vast majority of all the nitrogen fixed. There is a great diversity of organisms able to fix nitrogen, including the cyanobacteria, the phototrophs, *Bacillaceae*, *Enterobacteriaceae* and *Spirillaceae*, but the biochemistry of the process is remarkably similar in all the groups studied and the genes involved may be plasmid-borne and transmissible between groups. While being of particular evolutionary interest, this key physiological character is therefore of limited taxonomic value. Nitrogenase is oxygen-labile and so, paradoxically, these aerobic nitrogen fixers must carry out the process in microaerophilic conditions, as *Rhizobium* species do in root nodules, or have very high rates of respiration, as *Azotobacter* species do, to protect the enzyme from exposure to molecular oxygen.

For many years the aerobic, Gram-negative nitrogen fixers were classified in two families. *Azotobacter*, *Azotomonas*, *Beijerinckia* and *Derxia*, which are genera of free-living types, were placed in *Azotobacteraceae*, and the symbiotic species that formed *Rhizobium* were placed in the *Rhizobiaceae* along with the plant pathogens of *Agrobacterium*. Developments in molecular microbiology have confirmed some suspected relationships in the *Rhizobiaceae*, but indicated that *Azotobacteraceae* was not a natural family.

Azotobacteraceae

Four genera, *Azotobacter*, *Azomonas*, *Beijerinckia* and *Derxia*, were recognized within this family and their separation, initially based upon a few phenotypic characters, was later supported by mol% G+C, as shown in Table 9.3. In the light of a numerical taxonomic analysis however, two further, monospecific, genera were proposed to accommodate the atypical species *Azotobacter paspali*, which is loosely symbiotic with a certain grass, and *Azomonas macrocytogenes* which showed higher similarity to each other and then to *Azomonas* than they did to *Azotobacter* (Fig. 9.4). The first of these was not supported by nucleic acid studies (which showed that the rRNA cistrons of *Azotobacter paspali* and the other *Azotobacter* species are almost identical) or by immunoelectrophoresis, and the genus appears to be quite a tight one. It was also found that *Azomonas macrocytogenes* was only as distant from the two other species of the genus as they were from each other (Fig. 9.5). The genus is evidently fairly heterogeneous (its mol% G+C range for three species is wider than that for the six species of *Azotobacter*), but it was concluded that the creation of three monospecific genera was undesirable.

Table 9.3 Characters of some free-living, nitrogen-fixing, Gram-negative bacteria

Genus	Cellular morphology	Cyst* formation	Catalase	Autotrophic use of H_2 to fix N_2	mol% G+C
Azotobacter	Large, pleomorphic, ovoid rods	+	+	–	63–68
Azomonas	Large, pleomorphic, ovoid rods	–	+	–	52–59
Beijerinckia	Rods with polar PHB† granules	–	+	–	55–61
Derxia	Pleomorphic rods, beaded by PHB granules with age	–	–	+	69–73

*Resting form comprising central body, similar to vegetative cell, within a cyst coat.
† Poly-β-hydroxybutyrate.

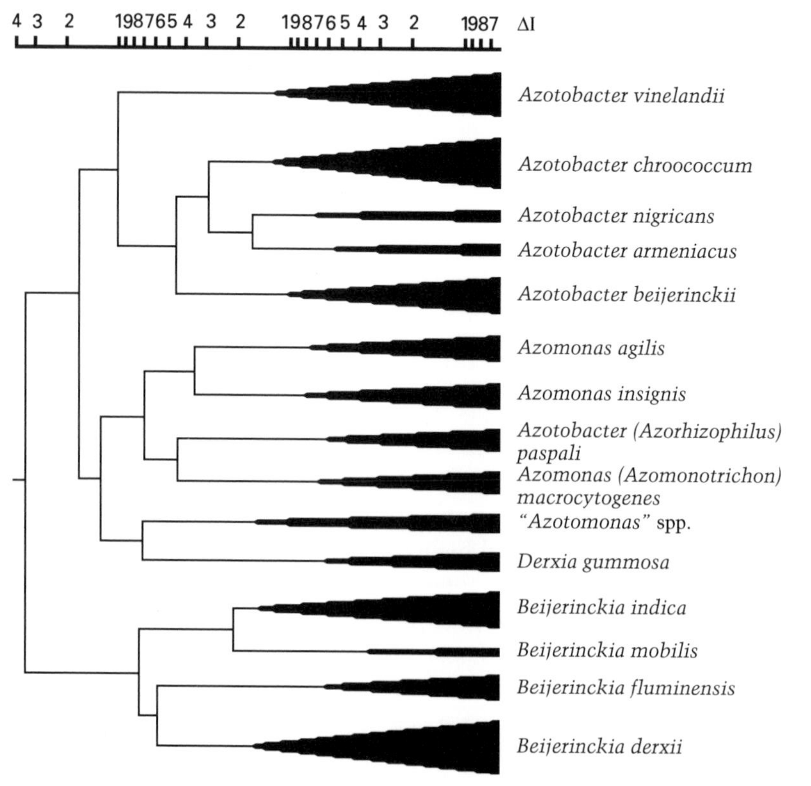

Fig. 9.4 A hierarchical classification of *Azotobacteraceae*. The simplified phenogram has been drawn showing information gain on an arbitrary, logarithmic scale. Adapted from Thompson & Skerman (1979).

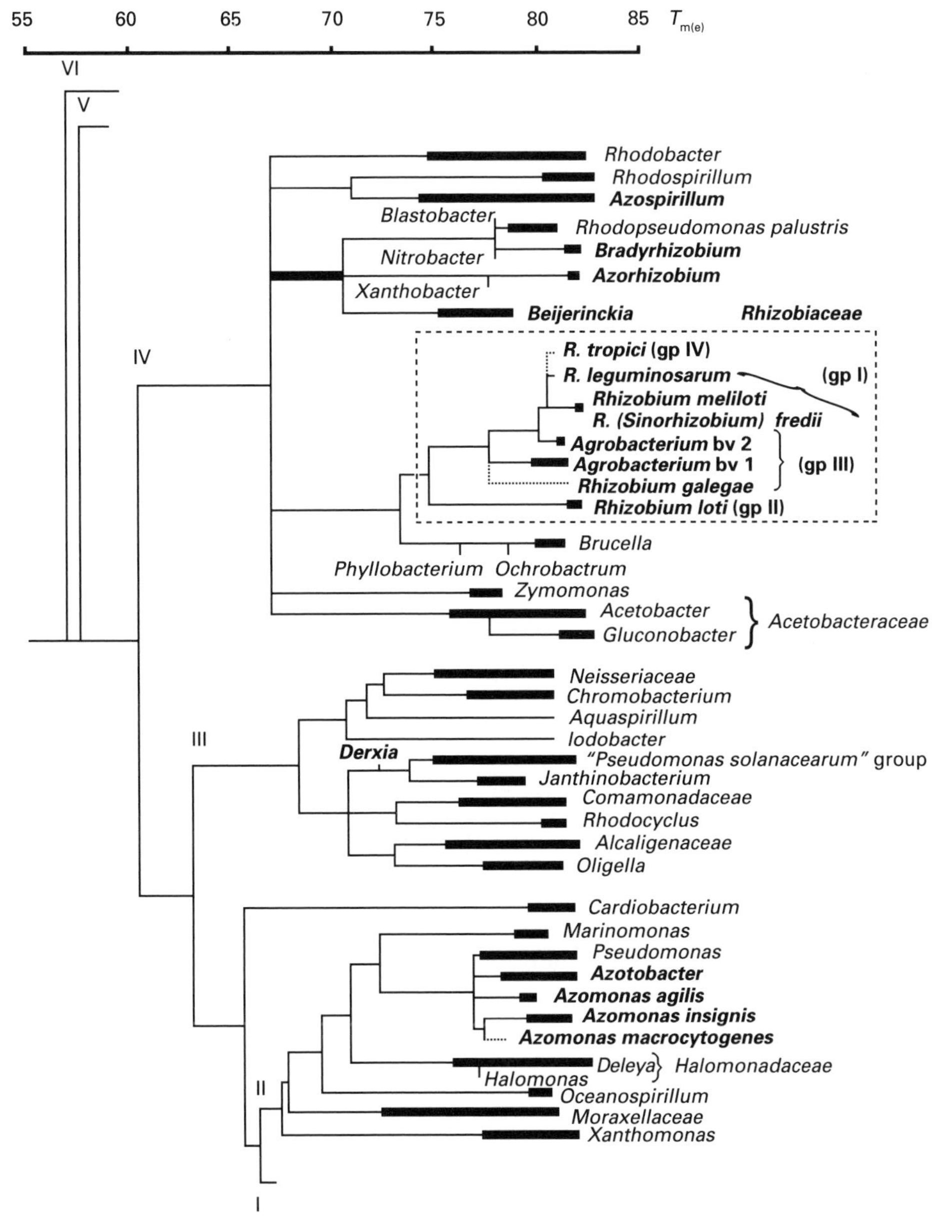

Fig. 9.5 Locations of genera of nitrogen-fixing bacteria and the agrobacteria, all indicated in bold print, in a rRNA cistron similarity tree.

Furthermore, it is clear from rRNA analysis that whereas *Azotobacter* and *Azomonas* are members of superfamily II (proteobacterial subclass γ) and are quite closely related to each other and to the true pseudomonads, *Beijerinckia* and *Derxia* do not belong to the family and are members of superfamilies IV and III (α and β subclasses) respectively. The numerical taxonomy findings are only in partial agreement, with *Beijerinckia* well separated but *Derxia* showing higher similarity to *Azomonas* than *Azomonas* does to *Azotobacter* (Fig. 9.4).

Rhizobiaceae

The main feature of this family, as previously defined, was that nearly all members cause hypertrophy of plant cortical cells. *Bradyrhizobium* and *Rhizobium* species infect the roots of leguminous plants and cause the production of nodules, within which they occur as pleomorphic forms, *Azorhizobium* nodulates stems as well, and all fix nitrogen within these nodules, and *Phyllobacterium* species nodulate the leaves of certain plants; *Agrobacterium* species, however, cause stem and root tumours and do not fix nitrogen. A close relationship between *Rhizobium* and *Agrobacterium* has been recognized for many years, none the less, but emphasis had been placed upon their so-called degenerate peritrichous flagellation (a combination of single polar and one or more peritrichous flagella) and even *Chromobacterium* was, for a time, included in the family on this basis.

The members of this group belong to two branches of superfamily IV (Fig. 9.5) which is the equivalent of the α subclass of the *Proteobacteria*. A point of special interest is the close relationship between *Rhodopseudomonas palustris* and *Bradyrhizobium*, as phototrophic, symbiotic nitrogen fixers belonging to the latter genus have been described; all fall within the same cluster and it has been suggested that they should be classified in *Rhodopseudomonas*, which has priority. *Azorhizobium* and *Beijerinckia* belong to the same branch, which is quite separate from those containing *Azospirillum* or *Agrobacterium* and *Rhizobium*; the latter group also contains *Aminobacter* (*Pseudomonas*) *aminovorans* (Fig. 9.3) and represents a more restricted *Rhizobiaceae*. Also of particular interest are the closeness of *Brucella* and (according to 16S rRNA sequencing) the rickettsias, as represented by *Rochalimaea quintana*, to *Agrobacterium* and *Rhizobium*, as all of these organisms show special relationships with eukaryotic cells.

Rhizobium. Early attempts at classifying these organisms defined species according to the groups of leguminous plants infected (representing a tiny percentage of the 19 700 species now known), but this approach, which is still used to define biovars, reflected little of the genome and gave way to more conventional methods as anomalous cross-infections were increasingly observed and it became recognized that some characters, such as host relationships, are plasmid-mediated. Following division of the genus into two groups according to rate of growth on yeast extract media, flagellation, mol% G+C, and host range, the slow-growing strains were accommodated in the new genus *Bradyrhizobium*. *Bradyrhizobium* itself is heterogeneous and is divisible into at least two groups by DNA relatedness and gene probing, serology, and antibiotic resistance patterns.

Numerical taxonomy, two-dimensional PAGE of cellular proteins, cellular fatty acid patterns, compositions of extracellular gums, serology, bacteriophage typing, plasmid transfer of infectivity and nucleic acid studies have all contributed to the recognition of the seven present species of *Rhizobium*, and the lipid A component of the lipopolysaccharide also promises to be a

Table 9.4 Groups of root, stem and leaf nodulating bacteria

Rhizobium rRNA group, genera, species and biovars	Main host	Special characters	mol% G+C	Closest relatives of group
I				
R. leguminosarum bv. *viceae*	*Vicia, Pisum*	Fast-growing, nodulate roots	59–63	
R.leguminosarum bv. *trifolii*	*Trifolium*			
R. leguminosarum bv. *phaseoli*	*Phaseolus*			
R. meliloti	*Medicago*	Fast-growing, nodulates roots	62–63	*Agrobacterium* cluster/biovar 2
R. fredii [*Sinorhizobium*]	*Glycine*	Fast-growing, nodulates roots	60–64	
R. huakuii	*Astragalus*	Fast-growing, nodulates roots	59–64	
II				
R. loti	*Lotus*	Fast-growing, nodulates roots	59–64	Member of *Rhizobium*, *Agrobacterium* and *Brucella* branch
III				
R. galegae	*Galega*	Fast-growing, nodulates roots	63	*R. meliloti* and *Agrobacterium* cluster/biovar 1
IV				
R. tropici	*Phaseolus* and *Leucaena*	Fast-growing, nodulates roots		*Rhizobium* group I
Azorhizobium caulinodans	*Sesbania*	Fast-growing, nodulates roots and stems	66–68	*Xanthobacter* and *Rhodopseudomonas palustris*
Bradyrhizobium spp.	Various	Slow-growing, nodulate roots	61–65	*Rhodopseudomonas palustris* and *Nitrobacter*
Phyllobacterium spp.	*Myrsinaceae* and *Rubiaceae*	Nodulate leaves; N_2 fixation unknown	59–61	*Ochrobactrum* and *Brucella*, in *Agrobacterium* and *Brucella* branch

useful taxonomic marker. From one polyphasic study, which included most of the approaches mentioned above, it was concluded that *R. fredii* should be split into two species in the new genus *Sinorhizobium*, but comparisons of 16S rRNA genes do not support the recognition of a new genus.

Four genetic groups are indicated by DNA–rRNA hybridization and ribosomal gene sequence studies, as shown in Table 9.4 and Fig. 9.5. Groups I and III are more closely related to the agrobacteria than they are to *R. loti*, which represents group II, and it has frequently been suggested that *Rhizobium* and *Agrobacterium* should be amalgamated in part, but as the vast majority of legumes await study for nodule bacteria the taxonomy of the genera must remain in a state of flux. Other complications include the isolation of non-nodulating rhizobias from soil and of star-shaped-aggregate-forming marine organisms closely related to *Agrobacterium*. Meantime, the International Subcommittee for the Taxonomy of *Rhizobium* and *Agrobacterium* has proposed minimal standards including symbiotic, cultural, morphological and physiological traits as well as phylogenetic studies for the description of new species and genera of these bacteria.

Agrobacterium. This genus is divisible into several groups by numerical taxonomy, cellular fatty acids, protein electrophoresis, serology and DNA–DNA hybridization, but these do not correlate with the recognized species (that is to say those included in the *Approved Lists of Bacterial Names*), whose classification was largely based upon plant pathogenicity. Pathogenic strains causing crown galls in a wide variety of plants were placed in *A. tumefaciens*, which is regarded as the type species; strains causing cane galls on *Rubus* spp. and other plants were placed in *A. rubi*; those causing hairy root or woolly knot of members of *Rosaceae*, and stem tumours were accommodated in *A. rhizogenes*; and non-pathogens found in the rhizosphere form *A. radiobacter*.

Table 9.5 *Agrobacterium* biovars and species

Biovar or group	Plasmid type* Ti	Plasmid type* Ri	Lactose oxidized to 3-ketolactose	Alkali from malonate	Acid from erythritol	Species name in *Approved Lists*
1	+		+	–	–	*A. tumefaciens*
	–		+	–	–	*A. radiobacter*
2	+		–	+	+	*A. tumefaciens*
	–		–	+	+	*A. radiobacter*
		+	–	+	+	*A. rhizogenes*
3	+		–	+	–	*A. vitis*
A. rubi	+		–	+	+	*A. rubi*

*Ti = tumour inducing type; Ri, hairy root inducing type.

Subsequently, phenotypic tests, reactions to monoclonal antibody, and DNA hybridizations showed biovar 3 strains from grapevines to be distinct from the other species and the new species *A. vitis* was proposed for them. The four clusters or biovars indicated by modern methods, and some of their properties, are shown in Table 9.5. *Agrobacterium* phytopathogenicity is plasmid-mediated, and it can be seen from the table that for *A. tumefaciens* the loss of the Ti plasmid would necessitate a change of name to *A. radiobacter*. This is most unsatisfactory, especially for a type species, and a change in nomenclature is needed, perhaps with biovar 1 containing *A. radiobacter*, as the type species, and *A. radiobacter* pathovar *tumefaciens* containing the tumorigenic strains.

Neisseriaceae, *Moraxella* and relatives

In the seventh edition of *Bergey's Manual* (1957) the family *Neisseriaceae* comprised two genera of parasitic, Gram-negative, non-motile cocci and the emphasis was on morphology as *Neisseria* contained aerobes and facultative anaerobes and *Veillonella*, which was subsequently removed to the new family *Veillonellaceae*, contained strict anaerobes. By the eighth edition of *Bergey's Manual* the family had been expanded to include two genera of coccoid rods, *Acinetobacter* and *Moraxella*, which had been classified in the *Achromobacteraceae* and *Brucellaceae* respectively, and the new genus *Branhamella* which accommodated *N. catarrhalis* and the cocci known as false neisseriae.

This last development and many subsequent changes in the taxonomy of this group resulted largely from studies of genetic affinity: a combination of transformation, and nucleic acid base compositions and homologies. Bøvre and colleagues at Oslo measured transformation compatibilities using streptomycin resistance as a marker, and determined G+C ratios and pulse-labelled (i.e. messenger) RNA–DNA homologies; RNase resistant mRNA–DNA hybrids test for closer relationships than do DNA–rRNA and some DNA–DNA pairing methods. They found that affinities between some species of *Moraxella* (rods) and *Branhamella* (cocci) equalled or exceeded those within the genera (Table 9.6) and proposed, as a compromise, that these two genera should become sub-genera of *Moraxella*. They also created the new genus *Kingella* to accommodate *M. kingae* which showed no affinity to *Moraxella*, and a little affinity to *Neisseria* (Table 9.6), yet was alone in possessing the enzymes nucleoside deoxyribosyltransferase and thymidine phosphorylase. The G+C ratios of the family overlap throughout the range 38–55 mol% and so are of limited value but patterns of cellular fatty acids, all straight-chained, show some correlation with genetic affinities, and together they have been used for identification and the selection of traditional tests for routine purposes.

The *Neisseriaceae* remained in a state of flux, however. Establishing the boundaries of the taxa at all levels was a problem and, as shown below, phylogenetic studies indicate that it was not a natural family (Fig. 9.6); it is, none the less, convenient to consider the genera together.

Neisseria

The members of this genus are human and animal parasites and typically occur as pairs of cocci with adjacent sides flattened. The importance of cell shape in classification is, however, called into question by the inclusion of a rod-shaped organism, *N. elongata*, in the genus; this species is genetically somewhat heterogeneous, and three subspecies are recognized, but it does show a close relationship with the other traditional neisserias by genetic affinity (Table 9.6), fatty acid composition, and possession of the enzyme carbonic anhydrase.

Close relationships are the rule in this genus, and the separation of species can be difficult. The three pigmented species *N. flava*, *N. perflava* and *N. subflava*, for example, were separated on the basis of a few unreliable cultural and biochemical tests and this unsatisfactory state of affairs led to their being incorporated into the single species *N. subflava*. Although the dozen species are separable by DNA relatedness, in many cases there are few biochemical differences and yet distinction between the pathogenic and commensal species is of great medical importance. The species *N. gonorrhoeae*, *N. lactamica*, *N. meningitidis* and *N. polysaccharea* are particularly closely related according to nucleic acid

Table 9.6 Genetic affinities amongst the neisserias and moraxellas

		Affinity in percentage of homologous reaction					
		Genetic transformation (recipient)			Hybridization (source of mRNA)		
DNA source	mol% G+C	M. nonlique -faciens	B. catarrhalis	N. elongata	M. nonlique -faciens	B. catarrhalis	N. elongata
M. nonlique -faciens	42	100	0.007	0	100	8.4	0.4
M. lacunata	42	0.4		0	22.2	4.5	
M. bovis	42.5	0.2	0.004	0	13.1	2.4	
M. osloensis	43.5	>0	0.002	0	2.0		
B. catarrhalis	41	>0	100	0	6.5	100	
B. ovis	45	>0	0.001	0	7.1	2.9	
K. kingae	44.5	0	0	>0	0.4		
N. flava	47.5	0		2.1	0.8	0.5	11
N. elongata	53	0	0	100	0.5	0	100
A. lwoffii	43	0	>0		0.6		

A., Acinetobacter; B., Branhamella; K., Kingella; M., Moraxella; N., Neisseria.
>0, low affinity, sometimes more distinct in reciprocal reaction; 0, no transformants.
From Bøvre (1970).

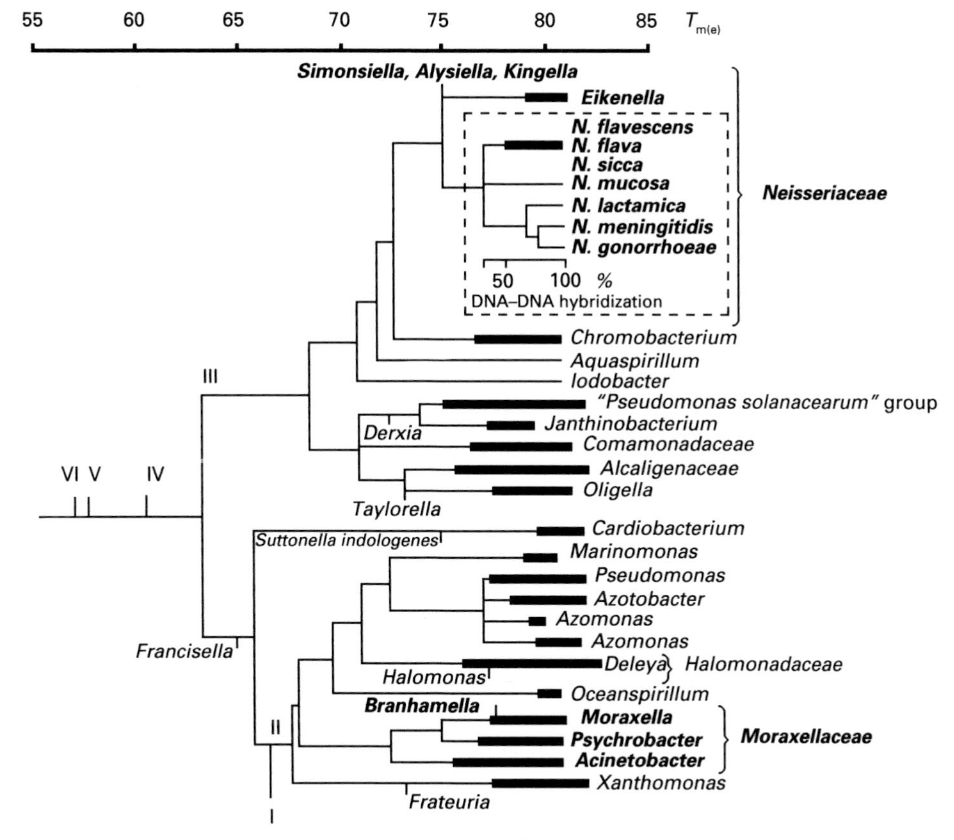

Fig. 9.6 Locations of the *Neisseriaceae* and *Moraxellaceae* in a rRNA cistron similarity tree. The *Neisseria* inset is derived from DNA–DNA hybridization data.

studies, with the first two showing 93% relatedness and the last one showing 69–71% relatedness to the group (Fig. 9.6); thus, they form a single genomic species with four subspecies that are differentiable by few traditional tests. *Neisseria cinerea* is closely related to this group and as a whole they show 0–46% relatedness to the other seven species.

When such important pathogens are involved clinical and practical considerations should override taxonomic ones, and so each member of the group retains its status as a separate species. Furthermore, the pathogenic species are separated by typing methods which are important for epidemiological purposes, serotyping and auxotyping (based upon nutritional requirements) being most used. The picture is complicated by the findings of a polyphasic study of some isolates from the eyes of Egyptian children. These organisms were not typable as gonococci and did not react with meningococcus antiserum, but had some phenotypic and genetic characters in common with these species, especially the latter, yet had different outer membrane protein patterns; on the basis of genetic affinity, however, they were classified as *N. gonorrhoeae* subsp. *kochii*.

Moraxella and Branhamella

Given the acceptance of a rod-shaped species in *Neisseria*, the separation of rods and cocci into the subgenera *Moraxella* and *Branhamella* seems unreasonable, but it was considered that the false neisseriae (*B. caviae*, *B. cuniculi* and *B. ovis*) warranted special designation. This arrangement gave rise to controversy about whether the false neisseriae are phenotypically and genetically close enough to *B. catarrhalis* to be included in the same subgenus. While it was generally agreed that the false neisserias did not belong to the genus *Neisseria*, the question of whether they belonged in the same family at all was in doubt; DNA–rRNA homologies now make it clear that they do not (Fig. 9.6) and that the family *Neisseriaceae* should include *Neisseria*, *Alysiella*, *Eikenella*, *Kingella* and *Simonsiella*, and belongs to superfamily III (proteobacterial subclass β) with *Chromobacterium* as its closest relative. It is also apparent that *Acinetobacter* is related to the *Moraxella–Branhamella* group, forming a sub-branch in superfamily II (proteobacterial subclass γ). Two proposals, one for the inclusion of *Moraxella* (and so *Branhamella*), *Acinetobacter* and *Psychrobacter* in a new family *Moraxellaceae*, and the other for the inclusion of *Branhamella* and *Moraxella* in a new family *Branhamaceae*, were published in the same number of the *IJSB*.

The organism "*M. urethralis*" was placed in the genus as a matter of convenience, but on the basis of a polyphasic study which included morphological, biochemical and carbon source assimilation tests, PAGE of cellular proteins, gas chromatography of cellular fatty acids, serology, genetic transformation and DNA–rRNA and DNA–DNA hybridization, the species was placed in the new genus *Oligella*, which is a member of the *Alcaligenes–Bordetella* sub-branch of superfamily III.

Acinetobacter

These plump, coccoid rods are ubiquitous and many independent isolations led to the use of a wide variety of names including *Achromobacter*, *Acinetobacter*, *Alcaligenes*, *Cytophaga*, *Diplococcus*, *Mima* and *Neisseria*. The first of these now contains oxidase-positive organisms, whereas *Acinetobacter* contains oxidase-negative strains — a feature which distinguished it from other members of the *Neisseriaceae*. Generally speaking, the acinetobacters have no special characters which distinguish them from other, similar bacteria and they are identified by a series of negative characters.

According to genetic transformation studies with nutritional markers *Acinetobacter* strains are quite closely related, and their range of 38–47 mol% G+C, taken with DNA–rRNA hybridization and studies of key enzymes in aromatic amino acid biosynthesis, supports the notion that the line representing this genus branched quite early during the development of the γ subclass of the *Proteobacteria* and that considerable evolution of individual strains has occurred. The genus has been divided into two groups using nutritional tests and according to whether acid is produced from carbohydrates, but the arrangements resulting from these two approaches do not correspond. Consequently, in *Bergey's Manual* (1984) only one species, *A. calcoaceticus*, was recognized, it being considered that there were no clear phenotypic criteria for establishing more. Subsequently, it has proved possible to separate some or all of the dozen genomic species using biochemical, growth temperature, and carbon source assimilation tests, and electrophoresis of outer membrane proteins, dehydrogenases and esterases. Also, it is possible to distinguish biovars, bacteriocin types and bacteriophage types within several of the species, and these are of epidemiological value, as some strains are opportunistically pathogenic.

The recently described radiation resistant *Acinetobacter* strains have been shown by DNA–DNA hybridization and other characters to form a group separate from the other members of the genus.

Legionellaceae

This family was established in 1979 to accommodate the genus *Legionella*, with its single species *L. pneumophila*, which was founded on the organism responsible for the 1976 outbreak of respiratory illness that occurred in Philadelphia and became known as Legionnaires' disease. A related organism had been isolated as early as 1943, but it and other isolates obtained in 1947 and 1959 were regarded as rickettsias as they were only cultivable by methods used for those organisms. The establishment of a one-member family reflected the apparent uniqueness of the Legionnaires' disease organism. It requires L-cysteine and iron salts for growth, possesses predominantly branched-chain fatty acids, uses amino acids as carbon and energy sources, and does not attack carbohydrates. Relationships with flavobacteria, which also produce quite large amounts of branched-chain fatty acids, and with other organisms showing some phenotypic similarities were excluded by DNA–DNA hybridization.

As the foremost taxonomist of this group, Don Brenner, stated in *Bergey's Manual* (1984) 'our knowledge of *Legionella* is in its infancy'; it still is. None the less, taxonomic development has been explosive; in less than 15 years 34 species with three subspecies and 53 serogroups, and at least five possible new species have been recognized. The medical importance of such opportunistic pathogens (16 species having been implicated in human disease) is not the least of the reasons for this rapid progress, but the recognition of so many species is largely owing to the methods that have been used in classifying the group.

The characters already noted are important in recognition at the genus level; Gram-negative, pleomorphic organisms requiring iron salts and cysteine for growth may be confirmed as legionellae by a few biochemical tests and their fatty acid patterns. There are relatively few phenotypic tests of value, however, and as a result of the medical importance of the organisms the division of the genus has been driven by the need for identification of strains in body fluids and from the environment, and species have been defined by serological methods much as with *Salmonella*. Such an approach is not entirely satisfactory as many species have been described on the basis of one strain, or a few strains from one source (though it is only fair to add that, for several of these, typical strains were subsequently isolated from other parts of the world). Also, identification may be restricted to reference laboratories, and strains not reacting with any of the polyclonal sera must represent new species or serogroups within a species: this must be addressed by DNA–DNA relatedness studies. However, numerical taxonomy of *L. pneumophila* with immunofluorescent staining using monoclonal antibodies has revealed phenons that largely correspond to those established using polyclonal sera and DNA studies.

A newly recognized bacterial group presents a great opportunity for using nucleic acid studies to produce a classification unhampered by previous arrangements, but certain difficulties have arisen. *Legionella* species fulfil the widely used definition of genomic species (see Chapter 3) but there are no satisfactory genetic definitions for higher taxonomic levels. It has been suggested that species of one genus should show 40–60% relatedness, but members of several well-established genera fall far short of this, as do most *Legionella* species, with figures of 25% or less. On this basis the creation of several families and many monospecific genera might be advocated. Indeed, a proposal to split the family into the three genera *Legionella*, *Fluoribacter* and *Tatlockia* was made at a time when only a handful of species had been described, some on the basis of very few strains. The rationale was that *T. micdadei* showed less than 10% relatedness to other species and that *F. bozemanii*, *F. dumoffii* and *F. gormanii* showed higher relatedness to each other than they did to the other species. Such a division was further supported by a serotaxonomic study (in which antigenic homology was expressed as the ratio of cross-reactive to total antigens analysed by crossed immunoelectrophoresis) and by subsequent 16S rRNA sequence comparisons and total carbohydrate patterns, on the basis of which the transfer of *L. maceachernii* to *Tatlockia* was proposed. It may be prudent to take a less hasty approach to the taxonomy of such a recently recognized group of bacteria, and from a practical point of view it is more convenient to emphasize the phenotypic similarities of the organisms and to lump rather than split.

Another phylogenetic study using 16S rRNA sequencing did not, however, support division of the family into three genera, finding that it represents a monophyletic group within the γ subclass of the *Proteobacteria* (superfamily I) (Fig. 9.7). None the less, it is apparent that the relationship between *L. pneumophila* and *L.* (*Tatlockia*) *micdadei* is not as close as that between several genera in the *Enterobacteriaceae*. It is of particular interest that the closest relatives of the family are the rickettsia-like *Coxiella* and *Wolbachia*, as all have intracellular parasitism of eukaryotes in common but represent a different line of proteobacterial descent from the true rickettsias and rhizobias, which belong to the α subclass (superfamily IV) (Fig. 9.7).

Some phenotypic characters such as colony fluorescence, browning on media containing diaminobenzoic acids or tyrosine, hippurate hydrolysis, and tests for lipase, protease and peroxidase are proving useful for

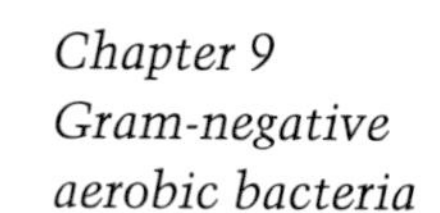

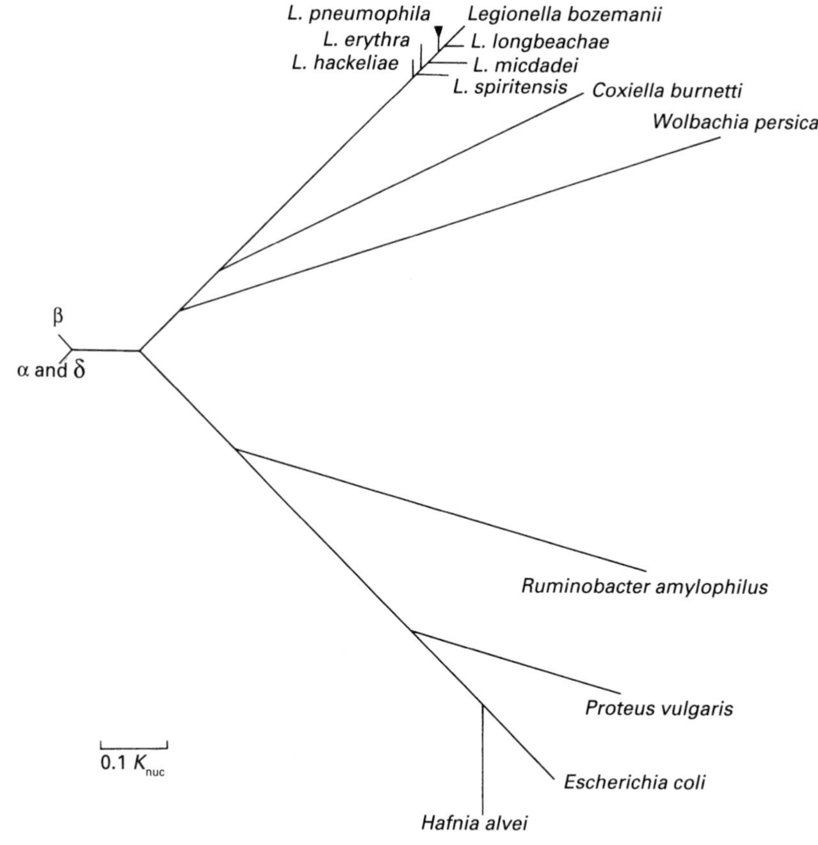

Fig. 9.7 Phylogenetic tree of some members of the γ subclass of the *Proteobacteria*, based upon 16S rRNA sequences, and showing relationships within the family *Legionellaceae*. Adapted from Fry *et al.* (1991).

separating the commoner, well-represented *Legionella* species and schemes have been devised for use in the routine laboratory. The reliability of such tests for the poorly represented, rarer species is unknown.

Other approaches, largely restricted to reference and research laboratories, include cellular fatty acid, isoprenoid quinone and sugar patterns, protein electrophoresis, dot hybridization (in which photobiotin-labelled test DNA hybridization with one of a selection of reference DNA dots on a filter is indicated by a colour reaction), rRNA gene restriction patterns and the use of nucleic acid probes. In one study, for example, capillary GLC revealed three major fatty acid groups (Fig. 9.8) and reverse-phase HPLC allowed five ubiquinone groups to be recognized among 182 strains representing 23 species (Table 9.7); qualitative and quantitative differences made further separation possible. However, another study using reversed-phase high-performance TLC found quite different ubiquinone groups. Most such methods were aimed at distinguishing between the currently recognized serogroups (i.e. species) as alternative approaches to identification but they will also be most valuable in polyphasic taxonomic studies of the family; especially because of the low levels of genetic relatedness between species.

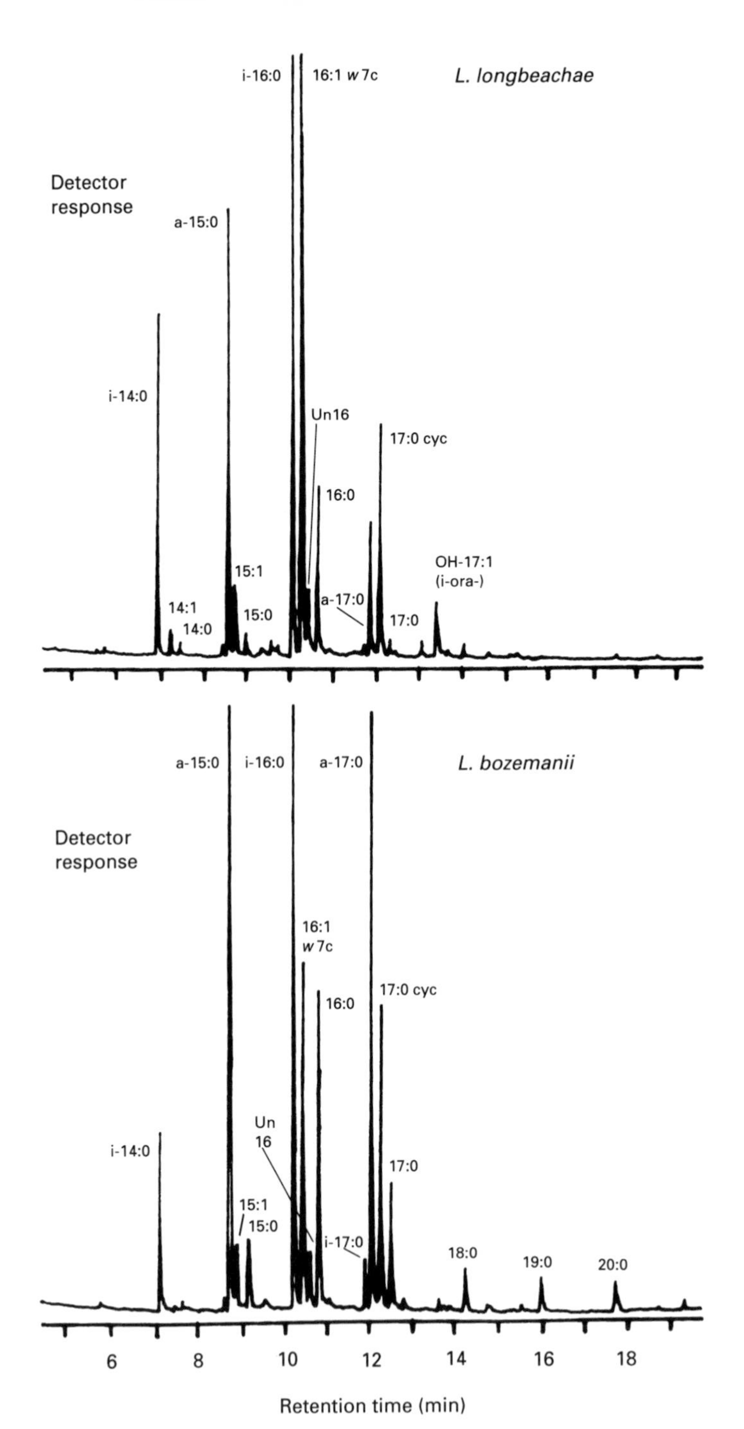

Fig. 9.8 Gas chromatograms of methylated cellular fatty acids of *Legionella longbeachae* and *L. bozemanii*, representing fatty acid groups 16C (containing 14-methylpentadecanoic [$C_{i\text{-}16:0}$] and hexadecanoic [$C_{16:1}$] acids as major components) and A15 (containing 12-methyltetradecanoic [$C_{a\text{-}15:0}$] acid as a major component) respectively. Explanation of fatty acid designations: i, or a, methyl branch at iso-, or anteiso-, carbon atom; number of carbon atoms: number of double bonds; *w*7c, double-bond position from hydrocarbon end of *cis* isomer; Un 16, unidentified 16-carbon acid; cyc, cyclopropane fatty acid; OH, hydroxy group.

Table 9.7 Ubiquinone groups of some *Legionella* species

Ubiquinone group* and species	Ubiquinone content†					
	Q9	Q10	Q11	Q12	Q13	Q14
A						
L. pneumophila			++	++++	++	
L. erythra			++	++++	++	
L. rubrilucens			++	++++	++	
B						
L. anisa	+++	++++	+++	+++	+	
L. cherrii	++	+++	++++	++++	+	
L. gormanii	+++	++++	++++	++++		
L. santicrucis	++	++++	++++	++++	(+)	+
C						
L. oakridgensis	++	++++	++	(+)		
L. wadsworthii	++	++++	+++	+		
D						
L. hackeliae				++	++++	+
L. jamestowniensis		(+)	++	++++	++++	(+)
L. maceachernii				+	++++	+
L. spiritensis				+	++++	+
E						
L. feeleii				(+)	++++	+++
F						
L. adelaidensis‡	(+)	++	++++	++		

*Groups A to E from Lambert & Moss (1989).
† Relative amounts from ++++ (major component) to (+) (trace).
‡ New species described by Benson *et al.* (1991) and appearing to represent a new ubiquinone group.

Other genera

Of the many other genera of Gram-negative aerobes some are well defined and some poorly so. They include *Acetobacter*, *Alcaligenes*, *Alteromonas*, *Beijerinckia*, *Bordetella*, *Brucella*, *Derxia*, *Flavobacterium*, *Francisella*, *Gluconobacter*, *Halomonas*, *Janthinobacterium*, *Lampropedia*, *Methylococcus*, *Methylomonas*, *Thermus* and *Xanthobacter*, but although some of their interrelationships and relationships with other genera have been revealed by rRNA sequencing and hybridization with DNA, in most cases it has not been possible to assign them to families.

An exception is the family *Acetobacteraceae* which contains the two well-defined genera of acetic acid bacteria, *Acetobacter* and *Gluconobacter*. They were previously separated, *Gluconobacter* being placed in the *Pseudomonadaceae* on account of its polar flagella, but DNA–rRNA hybridization revealed a close evolutionary relationship in a distinct branch of the α subclass of the *Proteobacteria* (superfamily IV) (Fig. 9.5) and this correlates with their many phenotypic similarities; both produce ellipsoidal to rod-shaped cells which are found in flowers, fruits (which they may rot), and alcoholic beverages, and they oxidize ethanol to acetic acid.

Methylotrophs

This large and diverse group contains organisms that use single-carbon compounds as energy and carbon sources. They have attracted considerable interest in recent years as sources of single-cell protein and enzymes which degrade certain environmental pollutants, and the rapid growth of information on them has emphasized the shortcomings in their taxonomy. Methylotrophs may be broadly divided into groups according to whether they are obligate or facultative methane-oxidizers (methanotrophs), or obligate or facultative oxidizers of methanol, methylamine and other methane derivatives but not methane itself. Facultative methylotrophs are found in a variety of Gram-negative and Gram-positive genera and some are quite restricted in the range of multicarbon compounds utilized.

Polyphasic numerical taxonomic studies on Gram-negative methylotrophs have included such characters as formaldehyde assimilation pathways, intracytoplasmic membrane structures, membrane fatty acid and isoprenoid quinone profiles, and DNA base ratios in addition to the usual morphological, physiological and biochemical tests. Two major groups of obligate methanotrophs could be recognized and the obligate methylotrophs and restricted facultative methylotrophs appeared to represent two distinct taxa warranting generic status, but it was clear that nucleic acid studies were required before formal taxonomic proposals could be made. Sequence analyses of 5S and 16S rRNA have shown that these organisms are scattered among the α, β and γ subclasses of the *Proteobacteria* in clusters that correlate with those recognized by traditional methods, and several taxa have been proposed following polyphasic studies, but the picture remains complex and many more strains must be studied before a satisfactory, overall classification can be established.

Brucella

The current classification of this genus was largely devised before the advent of modern taxonomic methods. The morphological, cultural and biochemical similarities between "*Micrococcus melitensis*", the cause of Malta fever in humans and abortion in goats, and "*Bacterium abortus*", the cause of abortion in cattle, was noticed by Evans in 1918 and the genus *Brucella* was created in 1920. *In vivo* and *in vitro* differences between strains from cattle and pigs led to the porcine variety of *B. abortus* being raised to species level as *B. suis* in 1929. These are known as the three classical, or major, species and they have been divided into 15 biovars. Of the three new, or minor, species, *B. neotomae*, from the desert wood rat, was added in 1966, and *B. ovis* from sheep and *B. canis* from dogs were added in 1970. Subsequent studies, including numerical taxonomies and nucleic acid analyses, have not led to any major changes to this practically satisfactory scheme.

The six species show a mol% G+C range of 55–58 and the very high DNA homologies between the species, all above 90%, indicate that the genus represents a genomic species and a very tight one at that. There are, consistent differences between the species, however, and subdivisions of the group correlate with host specificities and are epidemiologically valuable. Notwithstanding this, biovar-to-biovar variation does occur within species and there is inconclusive evidence of between-species variation. The characteristics used to separate the biovars are sometimes just matters of degree, for example *B. abortus* biovars 3 and 6 are separated by the former usually requiring CO_2, always producing H_2S and growing on media containing 1:25 000 (w/v) thionine and the latter not requiring CO_2, rarely producing H_2S and being inhibited by the thionine; strains with intermediate results for thionine have also been reported. Strains which do not match any of the current divisions have also been isolated, and it is probable that further biovars will be described.

At a higher taxonomic level DNA–rRNA hybridizations show that the closest relatives of *Brucella* are the clinical isolates placed in the new taxon *Ochrobactrum anthropi*, followed by the leaf-nodulating *Phyllobacterium* species, while 16S rRNA sequence comparisons of a different selection of organisms found a rickettsia, *Rochalimaea quintana*, to be the nearest relative. This evolutionary branch links with the *Agrobacterium* and *Rhizobium* cluster of superfamily IV (proteobacterial subclass α which also contains the mitochondria) (Fig. 9.5) and so it is of interest that members of most of these genera are noted for having intracellular existences in their hosts.

Important species

As might be expected, this loose assemblage of bacteria shows a very wide diversity of activities. *Beijerinckia, Bradyrhizobium, Derxia, Rhizobium, Xanthobacter* and members of the *Azotobacteraceae* are of great natural and economic significance in the nitrogen cycle and the symbiotic nitrogen fixers are often encouraged in agriculture by the planting of leguminous green-manure crops. The natures of the relationships between the rhizobiae and the agrobacteria and their host plants have been the subjects of intensive study and the tumour-inducing plasmid of *A. tumefaciens* has been used in the creation of genetically engineered, transgenic plants.

Also of great agricultural and horticultural importance are the phytopathogenic species of *Agrobacterium, Pseudomonas* and *Xanthomonas* which infect numerous foodplants and wild and ornamental species and cause diseases such as blights of beans (*P. syringae* pv. *phaseolicola* and *X. phaseoli*) and soybeans (*P. syringae* pv. *glycinea*) in which toxins have been shown to be involved, leaf blight of rice (*X. oryzae* pv. *oryzae*), blights and spots of cereals, tomatoes, peppers and stonefruits (*P. syringae* and *X. campestris* pathovars), wilts of potato, tomato, banana (Moko disease) and groundnut ("*P. solanacearum*"), wildfire of tobacco (*P. syringae* pv. *tabaci*, which produces tabtoxin), cabbage black rot (*X. campestris* pv. *campestris*), soft rots of vegetables (*Pseudomonas* species), sour skin of onions ("*P. cepacia*"), crown gall of members of the rose family and other plants (*A. tumefaciens*, whose oncogenicity is plasmid-mediated), olive knot (*P. syringae* pv. *savastanoi*), and cankers of stone and pome fruits (*P. syringae* pv. *syringae* and *P. syringae* pv. *morsprunorum*) and of citrus (*X. citri*).

Bacterial plant diseases are generally difficult to control, but the non-oncogenic *A. radiobacter* strain 84 has been used worldwide with great success for the biological control of crown gall disease, as most strains of *A. tumefaciens* are sensitive to the bacteriocin, agrocin 84, that it produces. Also, various *Pseudomonas* species have been reported to show bacteristatic, fungistatic, mycolytic and carcinoma-static activities and some may prove to be of value in biological control.

Turning to associations with the animal kingdom, the range of important activities is just as wide. Although primarily environmental organisms, *Legionella* species can infect healthy as well as debilitated persons and cause the serious pneumonia called Legionnaires' disease, or the milder, 'flu'-like infection known as Pontiac fever. The solanacearum group of pseudomonads contains the obligate animal parasite "*P. mallei*", which causes glanders in equines and many other animals, and the closely related but free-living soil organism "*P. pseudomallei*" which causes melioidosis in an even wider range of animals; both diseases are transmissible to humans. Several other pseudomonads are important as opportunistic pathogens, particularly in debilitated hosts such as those with burns, cystic fibrosis, or neoplasms; *P. aeruginosa* is the commonest, followed by *X. maltophilia* and "*P. cepacia*". The extensive resistance of pseudomonads to antibiotics and antiseptics causes special problems. Species of *Achromobacter*, *Acinetobacter*, *Alcaligenes*, *Flavobacterium*, *Moraxella* and other, as yet unnamed, organisms are often referred to, along with *Pseudomonas* species, as 'glucose non-fermenting Gram-negative bacteria' (NFB); they are of increasing importance in hospital-acquired infections and are frequently antibiotic resistant.

Brucellosis, causing abortion in animals and undulant fever in humans, remains an economically devastating disease in the many countries that do not have eradication schemes. Whooping cough, meningococcal meningitis and gonorrhoea, caused by *Bordetella pertussis*, *Neisseria meningitidis* and *N. gonorrhoeae* respectively, continue to be diseases of major importance worldwide. Permanent sequelae include brain damage or, in the case of gonorrhoea, sterility, and large sums are spent on treatment and prophylaxis. *Neisseria elongata* subsp. *nitroreducens* is emerging as an opportunist causing endocarditis, bacteraemia and osteomyelitis.

In addition to fighting bacterial diseases of plants, animals and himself, and making profits and losses from the activities of the acetic acid bacteria, man is now attempting to harness the activities of organisms such as the xenobiotic-degrading pseudomonads and the methylotrophs in the control of environmental pollution.

Identification

Most of the organisms covered in this chapter can be readily identified to genus level using traditional tests for growth conditions and morphological, physiological and biochemical characters. Schemes for identification to species level vary in their effectiveness and sophistication according to the clinical or other importance of the organisms and the tidiness of their taxonomies.

Identification within the NFB group has a reputation for being difficult because the organisms are often inert in routine tests and so several commercial systems have been developed with the clinical laboratory in mind; the API 20NE kit, for example, uses a mixture of traditional biochemical characters and assimilation tests. Such kits may be limited by their databases and this can be a particular problem when attempting to identify environmental isolates. The clinical importance of *P. aeruginosa* and other species is emphasized by the descriptions of various chemotaxonomic approaches to identification and the existence of several typing schemes based on serology, bacteriophages, bacteriocins and enzyme profiles.

For plant pathogens identification to genus and species level is usually straightforward using traditional methods and details of the host and its symptoms; it is not so easy with environmental isolates. Specific strains of *Agrobacterium* have been identified by immunodiffusion and DNA restriction enzyme digest patterns. Pathovars are defined by their host ranges and few phenotypic characters are useful for distinguishing between them. Other typing methods may be of use: phages specific for *X. oryzae* have been valuable in the forecasting of rice leaf blight, for instance, and *Xanthomonas* pathovars can be separated by their DNA restriction endonuclease digest patterns.

Standard phenotypic tests are also the mainstay for the identification of environmental organisms, but in some cases their sources or cultural requirements may be most helpful. Thus, *Rhizobium* species are isolated from root nodules, free-living nitrogen-fixers grow on nitrogen-deficient media, and methanotrophy can be detected by growing the organism in a methane atmosphere and assaying methane by gas chromatography before and after incubation. Various serological methods have been developed for the symbiotic nitrogen fixers, including ELISA with monoclonal antibodies.

As already indicated, legionellae are recognized by their requirement for cysteine and iron salts. A colony blot assay using genus-specific monoclonal antibody for rapid recognition of legionellae on plates has been developed. For the commonest species, identification by serological methods such as immunofluorescence and latex agglutination is very convenient and various commercial kits are available, but such an approach is not satisfactory for the other species because of the restricted availability of the sera and cost of holding them, and new serovars are regularly reported. Although chemotaxonomic and genetic probe methods and restriction endonuclease analysis are very effective, they are largely restricted to reference laboratories, and while phenotypic test schemes have been developed to avoid these problems, fatty acid analysis remains the method of choice.

Brucella isolates can be divided into species and biovars by well-established methods using measurements of oxidative metabolism of carbohydrates and amino acids by respirometry or TLC, growth in the presence of dyes, tests for urease, CO_2 requirement and H_2S production, bacteriophage sensitivities and serology. These methods are generally slow, however, and slide agglutination tests are dogged by cross-reactions with other organisms. A more rapid and specific method uses monoclonal antibodies to perform ELISA of colonies blotted onto nitrocellulose disks. Isolation of *Brucella* species from clinical cases is often difficult and diagnosis commonly relies on serology, the interpretation of which can be complicated.

Given the worldwide epidemic of gonorrhoea and the continuing importance of meningococcal meningitis and its speedy diagnosis, it is not surprising that many kits for the rapid identification of pathogenic neisserias are marketed. Several of the kits will also identify members of related genera such as *Branhamella*, *Moraxella* and *Kingella*. The methods include miniaturized, rapid versions of traditional carbohydrate fermentation tests, fermentation tests for radiometric monitoring of evolved $^{14}CO_2$ from blood cultures, chromogenic enzyme tests, coagglutination tests, immunofluorescence, and PCR-assisted nucleic acid probing.

10 Gram-negative, facultatively and strictly anaerobic bacteria

The Enterobacteriaceae are made up of a series of interrelated bacterial types which do not lend themselves to sharp division into tribes or into groups. Nevertheless, the family is so large and unwieldy that it is desirable to divide it into groups for purposes of practical classification.

F. Kauffmann, 1951, *Enterobacteriaceae*

Introduction

This chapter falls into five sections, each covering a family which contains clinically important organisms and whose taxonomy has been subjected to rapid and radical change in recent years: the *Enterobacteriaceae*, the *Vibrionaceae*, the *Aeromonadaceae* and the *Pasteurellaceae*, all of which show a close evolutionary relationship and belong to subgroup three of the γ subclass of the *Proteobacteria* (superfamily I) (Fig. 10.1), and the *Bacteroidaceae* which has been something of a dumping-ground for Gram-negative anaerobic rods.

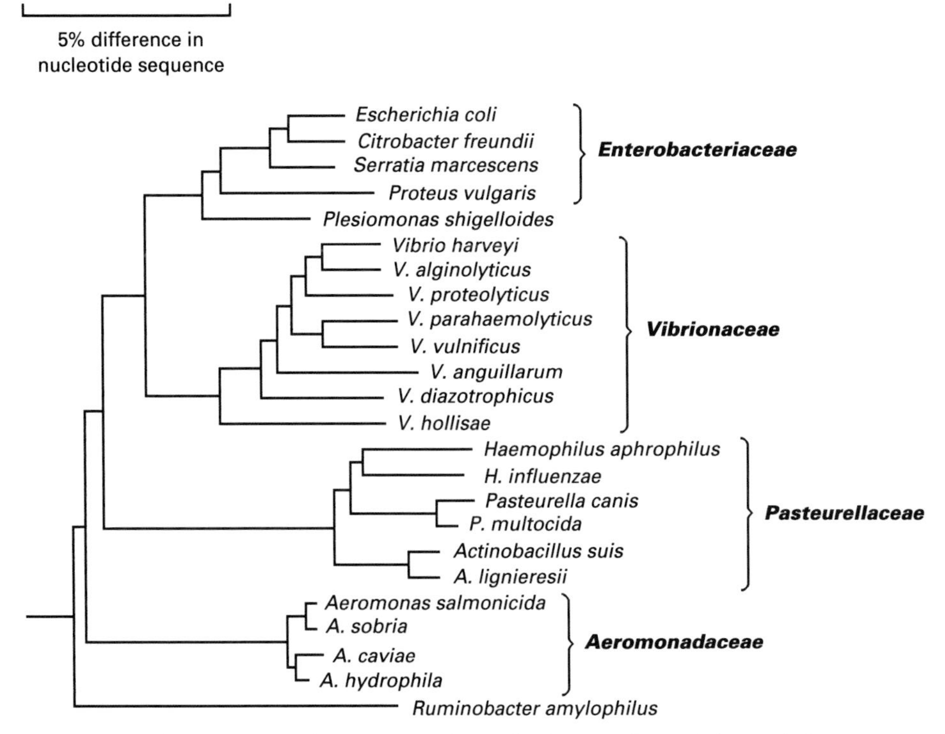

Fig. 10.1 Evolutionary tree for some members of subgroup three of the γ subclass of the *Proteobacteria*, based upon 16S rRNA sequence comparisons. Adapted from Dewhirst *et al.* (1992).

Enterobacteriaceae

The name of this family is misleading as it obscures its members' very wide diversity of habitats; these include water, soil, foodplants, ornamental plants and trees, and animals from insects to humans. The pathogenicity of several organisms, and the relatively early recognition and description of some of these, resulted in the establishment of genera and species which, as modern methods such as DNA relatedness show, might better be regarded as species and pathovars respectively. The most important organisms, medically speaking, have been subjected to extensive splitting, as seen in *Salmonella* with over 2000 serovars known by species names, whereas those of lower pathogenicity such as *Enterobacter* were lumped until recently. The medical importance of numerous species makes any major nomenclatural changes extremely difficult, if not impossible, to establish; obsolete names have a habit of surviving and the resultant confusion can have serious consequences. With this family therefore, the conflict between stability and a satisfactory taxonomy is particularly conspicuous.

With the *Enterobacteriaceae*, overclassification because of its members' great medical and veterinary importances has resulted in a tight family that is equivalent in taxonomic spread to a single genus elsewhere (Figs 10.1 & 10.2), *Bacillus* being the oft-quoted example. The family is not, therefore, phylogenetically deep and analysis of rRNA sequences and DNA–rRNA hybridizations are not always able to resolve the evolutionary picture at genus and species levels, so that it is necessary to compare more rapidly evolving cistrons. To date, such work has been based on too few groups to give anything more than a fragmentary picture, but comparative enzymology by studies of electrophoretic mobilities, regulation of biosynthetic pathways, assays of enzyme activities in these pathways, and amino acid sequencing are now allowing fine-tuning and producing pictures that are generally consistent with existing classifications. As an example, one study used 16S rRNA cataloguing data as a starting point to construct a dendrogram, and then used computer analysis of character states derived from comparisons of aromatic amino acid biosynthesis pathways to refine it; this revealed three 'enteroclusters' comprising: *Escherichia coli*, *Enterobacter*, *Klebsiella* and *Salmonella*; *Erwinia*, *Pantoea* and *Serratia*; and *Cedecea*, *Edwardsiella*, *Kluyvera*, *Morganella*, *Proteus* and *Yersinia*.

Enzyme electrophoresis has been particularly widely used, alone and as part of polyphasic studies. On the one hand it has allowed the recognition of a globally distributed clone of *Salmonella typhi*, and a clone restricted to Africa, and on the other it has been shown that, although *E. coli* may show great conservation of plasmid types across the world over many years, enzyme electrophoretic types may show some correlation with the ecological niches of the strains. The pace of change in the *Enterobacteriaceae*, which now contains 32 genera and over 100 species, is shown in Table 10.1. So many new groups may seem to indicate a splitting mentality, and such an approach would be in the best tradition of enterobacterial classification, but in most cases the new taxa are merely names put to previously

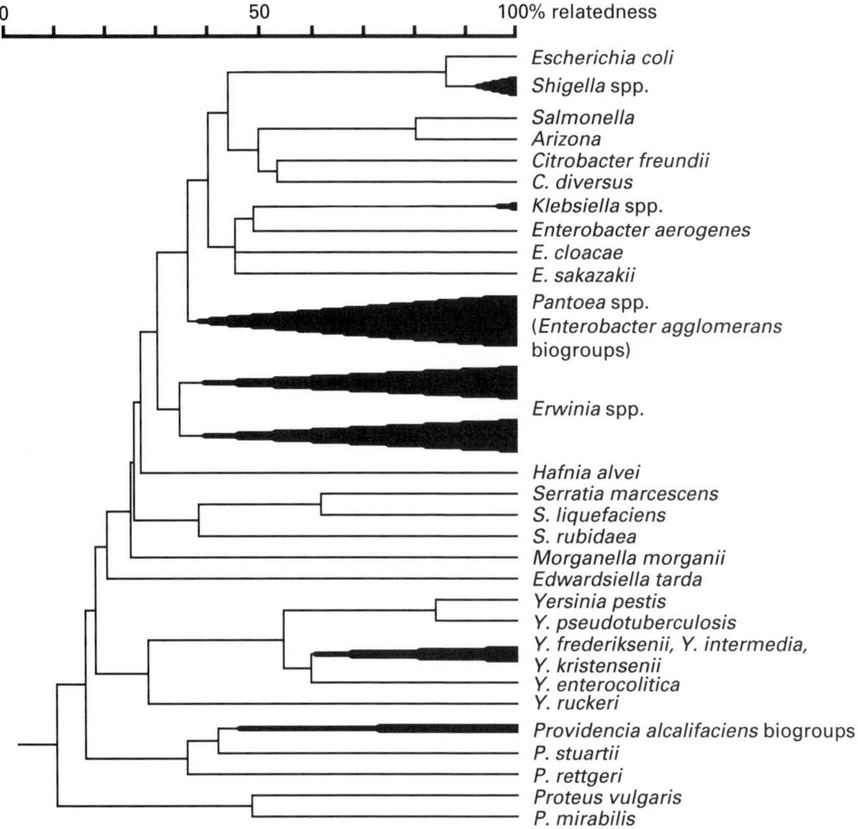

Fig. 10.2 Dendrogram based on DNA relatedness groupings, and assuming a common ancestor, indicating the evolutionary relationships of members of the *Enterobacteriaceae*. Adapted from Brenner (1984) *Bergey's Manual of Systematic Bacteriology*, Volume 1, p. 410.

recognized groups of atypical organisms that awaited adequate characterization and computer analysis.

Escherichia and Shigella

For many years the type genus of the family, *Escherichia*, was monospecific and it was believed to be a harmless gut commensal, with only *Shigella* and *Salmonella* recognized as gastrointestinal pathogens. The division of these groups was originally based upon their pathogenicities, and the biochemical tests that were devised to differentiate them were also used for their definition; *E. coli* was motile and produced acid and gas from lactose, whereas *Shigella* species were non-motile, and rarely produced acid from lactose or gas from any carbohydrate. However, it is now appreciated that some *E. coli* strains are important causes of diarrhoeal and other illnesses, and non-motile, lactose-negative, anaerogenic *E. coli* (the Alkalescens–Dispar group) have long been recognized, as have aerogenic *Shigella* species. One of the latter, *S. boydii* serotype 13, is also separable from the rest of the group by esterase electrophoresis and DNA relatedness and it has been suggested that

Table 10.1 Members of *Enterobacteriaceae*

Year of proposal	Name	Sources	Number of species		
			In *Bergey's Manual**	In 1993	Of clinical importance
1823	*Serratia*	Clinical† and environment	6	10	1
1885	*Klebsiella*	Clinical and environment	4	6	4
1885	*Proteus*	Clinical, animals and environment	3	4	3
1900	*Salmonella*	Clinical and animals	1(6)	2(6)‡	1(5)
1919	*Escherichia*	Clinical and animals	2	5	1+
1919	*Shigella*	Faeces	4	4	4
1920	*Erwinia*	Plants	15	16	0
1932	*Citrobacter*	Clinical and environment	3	3	3
1943	*Morganella*	Clinical and animals	1	1	1
1944	*Yersinia*	Clinical, animals and environment	7	11	3
1954	*Hafnia*	Clinical, animals and environment	1	1	1
1960	*Enterobacter*	Clinical, animals and environment	5	12	4
1962	*Providencia*	Clinical	3	5	3
1963	*Obesum-bacterium*	Breweries	1	1	0
1965	*Edwardsiella*	Animals and clinical	3	3	1+
1979	*Rahnella*	Water and clinical	1	1	0
1979	*Xenorhabdus*	Nematodes and insects	2	4	0
1981	*Buttiauxiella*	Water	1	1	0
1981	*Cedecea*	Clinical	3+	3+	?
1981	*Kluyvera*	Environment and clinical	2	2	2?
1981	*Tatumella*	Clinical	1	1	1
1983	*Ewingella*	Clinical	0	1	1
1984	*Moellerella*	Faeces	0	1	?
1985	*Budvicia*	Water	0	1	0
1985	*Leminorella*	Faeces and urine	0	2	?
1985	*Yokenella*	Clinical	0	1	?

continued on p. 147

Table 10.1 *continued*

Year of proposal	Name	Sources	Number of species: In *Bergey's Manual**	In 1993	Of clinical importance
1986	*Leclercia*	Clinical and environment	0	1	?
1988	*Pragia*	Water	0	1	0
1989	*Pantoea*	Environment,fruit, water and clinical	0	7	1+
1991	*Trabulsiella*	Environment and clinical	0	1	0
1991	*Arsenophonus*	Parasitic wasp	0	1	0
1993	*Photorhabdus*	Nematodes and insects	0	1	0

* *Bergey's Manual of Systematic Bacteriology,* Volume 1 (1984).
† Meaning specimens from diseased and healthy persons.
‡ Taxonomy of the genus is unsatisfactory; this arrangement recognizes two species and six subspecies.

it represents a new species. A numerical taxonomy using traditional characters separated these groups and found Alkalescens–Dispar and *Providencia* to be the closest relatives of the four *Shigella* species.

However, modern methods such as enzyme electrophoresis and nucleic acid studies indicate a particularly close relationship between *Shigella* and *E. coli*; the shigellas show 70–100% DNA relatedness to *E. coli* and are merely metabolically inactive biogroups of the latter. Because of their clinical importance and familiarity, a compromise must therefore be made, since the single species that these organisms represent is known by two generic names and over five species names.

The picture is further complicated by the proposal of four new species of *Escherichia*: *E. blattae*, *E. fergusonii*, *E. hermannii* and *E. vulneris*, none of which are as closely related to *E. coli* as *Shigella* species are. Comparisons of glyceraldehyde-3-phosphate dehydrogenase and outer membrane protein 3A gene sequences indicated that while *E. coli* and *E. fergusonii* belong in the same genus, *E. hermanni* and *E. vulneris* are further distant than *Salmonella* and *Citrobacter*, with *E. vulneris* lying closest to *Klebsiella pneumoniae*. *Escherichia blattae* is almost as distant from *E. coli* as is *Serratia*. It had been suggested that since the long-neglected organism *Escherichia adecarboxylata* is yellow-pigmented (as, incidentally, are most strains of *E. hermannii* and *E. vulneris*) and negative in the routine decarboxylase tests, it belongs to the *Enterobacter agglomerans–Erwinia* group, but it has now been elevated to generic status as *Leclercia adecarboxylata* on the basis of phenotypic tests, protein electrophoresis, and low DNA homologies with other members of the family.

Salmonella

Owing to their clinical importance early isolates of *Salmonella* were given names such as *S. abortus-bovis*, *S. cholerae-suis*, *S. typhi-murium* and *S. typhi*, which indicated the nature of the disease and, usually, the animal involved. It soon became clear that this was unsatisfactory because the limitation of host and pathogenicity implied was misleading for those who expected a name to be more than just a label; *S. typhimurium*, for example, is better known as a cause of human gastroenteritis than of mouse typhoid. None the less, these species names are still widely used. Later nomenclature was based upon the geographic location of the first isolation, resulting in *S. guildford*, *S. bristol* and *S. glasgow* for example.

The recognition of individual serovars within the genus is most important as it permits the tracing of sources of food-borne infection outbreaks. The Kauffmann–White scheme, in which organisms are represented by the somatic, flagellar and capsular antigens of primary diagnostic value, is considered to be of overriding importance and has yielded over 2000 serovars. Further division, of serovars into phagovars, biovars and resistovars (to bacteriocins or antibiotics), makes precise epidemiological studies possible. However, the continued practice of giving names to the serovars pathogenic for warm-blooded animals implied species status, so that *Salmonella*, representing a species in the *Escherichia* group rather than a genus, was regarded as having some 1500 species.

From 1960 Kauffmann divided the genus into four subgenera on the basis of biochemical characters. A fifth subgenus has subsequently been recognized. These divisions correlated with the organisms' ecologies and subgenus III represented the organisms of the Arizona group, members of which are often lactose positive and which Ewing proposed be placed in the separate genus *Arizona* (Table 10.2). In 1973 Crosa and co-workers recognized five subgroups on the basis of DNA relatedness; these corresponded with the four subgenera, but subgenus III was split into two groups that correlated with phenotypic characters such as lactose fermentation and whether flagellar antigens were monophasic or diphasic.

Over the years various proposals had been made either to limit the genus to only three species (*S. typhi*, *S. choleraesuis* as type species, and a single further species to embrace all the other serovars), or to recognize the subgenera as species. In 1982 Le Minor and colleagues proposed the single species *S. choleraesuis* with six subspecies on the basis of a numerical taxonomy of carbon source utilization and other biochemical tests followed by DNA hybridizations of representative strains (Table 10.2). It was subsequently proposed in the light of a comparison of various metabolic enzymes by starch-gel electrophoresis and computer analysis, taken along with DNA relatedness, that *S. choleraesuis* subsp. *bongori* be given species status. It was also found that the two subgenus III subgroups are more closely related to other subgenera than they are to one another.

These are major taxonomic improvements, but the *Enterobacteriaceae* Subcommittee of the International Committee on Systematic Bacteriology

Table 10.2 Divisions of the genus *Salmonella*

Scheme	Proposed groups					
Subgenus	I	II	III	III	IV	V
DNA group of Crosa *et al.*	1	2	3	4	5	—
Genus according to Ewing	*Salmonella*	*Salmonella*	*Arizona*	*Arizona*	*Salmonella*	*Salmonella*
Subspecies* of Le Minor *et al.*	*choleraesuis*	*salamae*	*arizonae*	*diarizonae*	*houtenae*	*bongori*
Mainly from warm or cold blooded animals	Warm	Cold	Cold	Cold	Cold	Cold
Human pathogens	++++	+	+	+	+	?
Flagella†	Di	Di	Mono	Di	Mono	Mono

*A seventh subspecies, *S. choleraesuis* subsp. *indica*, has been described.
† Flagella antigens usually diphasic or monophasic.

considers the Kauffmann–White scheme to be of prime importance and that naming of the subgenus I (*S. choleraesuis* subsp. *choleraesuis*) serovars should continue. It has been proposed that the species name be changed to *S. enterica* to avoid confusion, but as it is inconvenient for the clinical bacteriologist to report *Salmonella enterica* subsp. *choleraesuis* serovar *typhimurium* such strains will routinely be referred to as S. *typhimurium* which will be perceived as a species. So, wherever the taxonomy of the genus leads, nomenclatural usage is unlikely to follow.

Enterobacter, Erwinia and Klebsiella

The type species of these three groups are quite distinct and readily recognized, but the boundaries of the genera have been obscured by the assortment of atypical strains and species that have been allocated to them over the years; as a result there are taxonomic and nomenclatural difficulties.

The major problem was the *Erwinia herbicola–Enterobacter agglomerans* group. The latter species was proposed by Ewing & Fife in 1972 to accommodate organisms previously assigned to *Erwinia* (partly because of their yellow pigment) under a variety of names which they believed to be synonymous: *E. herbicola*, *E. stewartii* and *E. uredovora*, for example. Biochemical tests showed the group to be heterogeneous, being divisible into 11 biovars, and closer to *Enterobacter* than *Erwinia*, but it was suggested that it might represent a new taxon. However, the type strains of the three erwinias differ and it was convenient for plant pathologists to continue to recognize these species, so that only medical bacteriologists tended to use the name *Enterobacter agglomerans*. Taxonomists, then, spoke of the *Erwinia herbicola–Enterobacter agglomerans* complex and subjected it to intensive study by numerical taxonomy, protein electrophoresis and DNA hybridization. The different approaches indicated between 13 and 23 phenons and identified areas for closer study, and subsequently the new genus *Pantoea* with the species *P. agglomerans* and *P. dispersa* was proposed on the basis of DNA relatedness and electrophoretic protein patterns. Both species contain strains previously called *Enterobacter agglomerans* and *Erwinia herbicola* and the former also contains the type strains of these species and that of *Erwinia milletiae*. Several other groups may represent additional species of *Pantoea* (for example, the type strains of *Erwinia ananas*, *E. stewartii* and *E. uredovora* showed between 39% and 56% homology to species of the new genus), but allocation of these and many other organisms awaits the study of further strains.

Other changes include the transfer of the atypical *Erwinia* species, *E. dissolvens* and *E. nimipressuralis* (both closely related to *Enterobacter cloacae* and of uncertain plant pathogenicity), to *Enterobacter*. Even with the removal of all these species, *Erwinia* remains a heterogeneous genus (Fig. 10.2). It has been suggested that the genus be limited to pathogens such as *E. amylovora* and that the nutritionally more versatile, soft rotting species such as *E. carotovora* be placed in a separate genus *Pectobacterium* but intermediate strains make delineation of the genera difficult.

Another proposal was the transfer of *Enterobacter aerogenes* to *Klebsiella* as it is equally or more closely related to members of that genus, according to biochemical tests and DNA relatedness studies (Fig. 10.2), than it is to *Enterobacter*. Such a move presents two difficulties: first, *Klebsiella* would have to be redefined to accommodate motile organisms; and second the name *K. aerogenes* has been widely used as a synonym for certain *K. pneumoniae* strains, but the alternative name suggested, *K. mobilis*, would be misleading given the existence of non-motile strains. This genus provides another example of taxonomy and nomenclatural usage parting company: *K. pneumoniae*, *K. ozaenae* and *K. rhinoscleromatis* are easily distinguished in the laboratory (as they show decreasing metabolic activity in the given order) and their medical importance supports their continued separation — but they really represent bio-sero-pathogroups of one genomic species and have been reclassified as subspecies of *K. pneumoniae*.

Yersinia

This genus was created to accommodate *Pasteurella pestis* and *P. pseudotuberculosis* which differed phenotypically and genotypically from other members of that genus; their original inclusion had been on the basis of microscopic morphology. Later *Y. enterocolitica* was added and subsequent isolations showed it to be part of a large group containing a homogeneous genomic species divisible into five biovars, and various *Y. enterocolitica*-like organisms and atypical biovars. Nucleic acid studies showed *Yersinia* to be a homogeneous taxon, with its species showing greater similarity to each other than to any other members of *Enterobacteriaceae* (with which it shares the common antigen), and have supported the recognition of seven new species. Three of these (*Y. frederiksenii*, *Y. intermedia* and *Y. kristensenii*) represented

Y. enterocolitica-like organisms and two (*Y. bercovieri* and *Y. mollaretii*) the atypical *Y. enterocolitica* biovars 3A and 3B. While not practically desirable, further subdivision is possible as *Y. frederiksenii* comprises three DNA relatedness groups that are phenotypically indistinguishable, and rDNA polymorphism analyses revealed between two (*Y. enterocolitica*) and five (*Y. frederiksenii*) groups, which correlated with enzyme electrophoresis patterns, within several species. While the breadth of the *Y. enterocolitica* group was becoming appreciated, the closeness of *Y. pestis* and *Y. pseudotuberculosis* was being recognized; these two species show 90% relatedness and constitute a single genomic species (Fig. 10.2), and it was proposed that *Y. pestis* should become a subspecies of *Y. pseudotuberculosis* for taxonomic, if not medical, purposes. This conflict of taxonomy and practical nomenclature brings a further problem because *Y. pestis* is the type species of the genus.

Vibrionaceae

This family was proposed in 1965 as a convenient grouping of organisms which differed from members of the *Enterobacteriaceae* by being, in the main, oxidase-positive and motile by polar rather than lateral flagella. As shown below, the two families are evolutionarily related and studies on enzyme regulation and nucleic acid sequencing have been particularly rewarding. The family presently comprises the four genera *Listonella*, *Photobacterium*, *Shewanella* and *Vibrio* whose members are primarily associated with aquatic habitats, and range from marine and freshwater saprophytes through to parasites, endosymbionts and pathogens of humans, fish, amphibians, shellfish and other invertebrates.

Vibrio

The number of species in this genus has greatly increased in recent years in the light of numerical taxonomies, immunological comparisons of the enzymes superoxide dismutase (SOD) and alkaline phosphatase, and nucleic acid studies. Five species were listed in the eighth edition of *Bergey's Manual* (1974), 27 were listed in the 1984 edition, and by 1993 there were 36 valid species. Of the 31 additions, less than half are species transferred from other genera; nine from *Beneckea* and one from *Lucibacterium*, with both of these genera losing nomenclatural standing, and one each from *Aeromonas* and *Photobacterium*. Two of the remainder were previously biovars of the established species *V. anguillarum* and *V. cholerae*. Several other biovars probably warrant species status on phenotypic and genotypic grounds, but await better representation. The rest are new species, and in several cases their inclusion has required redefinition of the genus, to include, for example, oxidase- and nitrate reduction-negative strains, and nitrogen fixers. Furthermore, several other groups of strains have been recognized and may warrant species status once adequately characterized.

The genus spans a range of 38–51 mol% G+C and its members are quite diverse in their ecologies so that no one sodium chloride concentration or incubation temperature is optimal for all species. None the less, numerical taxonomies based upon traditional phenotypic tests have made, and continue to make, valuable contributions towards understanding relationships within the genus and fully agree with the conclusions reached from nucleic acid and enzyme analyses. Other chemotaxonomic methods have been little used. Most studies have concentrated on collections of strains isolated from particular environments or sharing special properties, such as human pathogens, fish pathogens, nitrogen fixers and organisms from brackish water or sea water.

The relationships between *Vibrio* and other genera have been investigated using DNA–rRNA hybridization studies and immunological comparisons of SOD and glutamine synthetase. This work revealed a group of closely related marine species, but other species, particularly *V. gazogenes*, *V. metschnikovii*, *V. cholerae* and *V. costicola* lay at increasing distances from the marine group and did not form groups themselves. *Lucibacterium harveyi* and most *Beneckea* species lay within the marine group, the remainder being scattered nearby, and it was this distribution that led to these two genera losing nomenclatural standing (Fig. 10.3). There are relatively few data on the larger rRNA molecules for *Vibrionaceae*, but complete sequencing of 5S rRNA followed with cluster analysis of difference

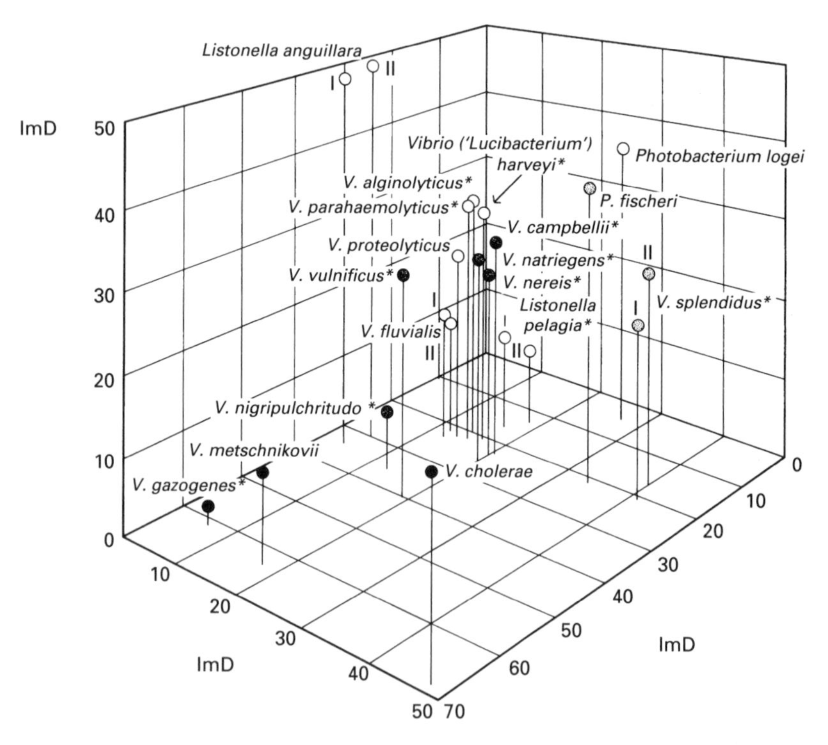

Fig. 10.3 A three-dimensional plot to show the immunological relationships of superoxide dismutases of *Vibrio* species and relatives, as represented by immunological distances (ImD) that are related to percentages of amino acid sequence difference. Organisms previously accommodated in the genus *Beneckea* are indicated by asterisks. Adapted from Baumann *et al.* (1984) *Bergey's Manual of Systematic Bacteriology*, Volume 1, p. 530.

matrices has been valuable in revealing relationships within the family and has resulted in its restructuring and the proposal of the two new genera, *Listonella* and *Shewanella*. It has also been suggested that several new *Vibrio* species properly belong in *Listonella*.

Subsequently, comparisons of 16S rRNA sequences of marine bacteria indicated that the family *Vibrionaceae* could be separated into six or more groups at the genus or family level:

1 *V. cholerae* and *V. mimicus*;

2 A group of 16 *Vibrio* species including *V. alginolyticus*, *V. gazogenes*, *V. harveyi*, *V. metschnikovii*, *V. natriegens* and *V. parahaemolyticus*, and *Listonella pelagia*;

3 *V. marinus* (to which *Shewanella putrefaciens* is related);

4 *V. costicola*, *V. hollisae* and *Listonella damsela*;

5 *P. fischeri*, *P. logei* and *Listonella anguillara*;

6 *Photobacterium*.

The main phenotypic characters used for classifying and identifying the vibrios show very limited correlation with the phylogenetic scheme, but hopefully this problem may be solved by polyphasic taxonomic studies.

Aeromonadaceae

On the basis of phenotypic properties the genus *Aeromonas* was previously allocated to the *Vibrionaceae*, but the evolutionary relationships implied by DNA–rRNA hybridization, 16S rRNA oligonucleotide cataloguing, and 5S rRNA sequencing studies showed that it is no more a member of that family than are the enterobacteria. The single-genus family *Aeromonadaceae* was therefore proposed, but it has not been universally accepted. *Plesiomonas* is likewise distinct but has not been proposed as a new family, and it appears to be closely related to the families *Aeromonadaceae* and *Enterobacteriaceae*.

Aeromonas

This genus was readily split into two divisions of species by phenotypic tests, but DNA homologies indicated 12 groups, and opinion was divided as to how many species or subspecies each division should contain; between the splitters and lumpers, in other words. The first phenotypic division includes non-motile, psychrophilic aeromonads of the species *A. salmonicida*. Four subspecies of *A. salmonicida* are distinguishable using biochemical tests, but as they show high DNA relatedness some authorities believe that they do not warrant subspecies status and should be regarded as biovars. Others consider that it is useful to split the species because typical strains from salmonid sources throughout the world form a very tight genetic grouping (*A. salmonicida* subsp. *salmonicida*), while some of the other subspecies, atypical strains from salmonid and non-salmonid sources, show quite high relatedness to *A. hydrophila*, which belongs to the second division of the genus.

This hydrophila division contains motile organisms capable of growing at 37°C. They have attracted much attention in recent years because of their role as opportunistic pathogens. Some workers believe that the group should be contained in the one readily identifiable species, *A. hydrophila*, but more recent numerical taxonomic studies, PAGE of whole-cell hydrolysates and [^{35}S] methionine radiolabelled proteins (radioPAGE), PAGE and isoelectric focusing of esterases and dehydrogenases, GLC of fatty acid methyl esters, and DNA relatedness supported the recognition of the three species *A. hydrophila*, *A. punctata* (replacing *A. caviae* by priority) and *A. sobria*. Furthermore, it was found that potential virulence factors are correlated with phenotypic characters of these three species. However, each contains several relatedness subgroups, and subgroups are also discernible by numerical taxonomy and radioPAGE. In consequence the division has been reclassified by DNA group to contain five new taxa (four species, one with two biogroups), with one more species representing a further DNA group, all of which are distinguishable by conventional tests.

Unfortunately, 16S rDNA sequence comparisons have indicated several relationships in disagreement with those implied by DNA hybridization studies. Strains of *A. hydrophila* and *A. media*, for example, showed low DNA relatedness but very high rDNA sequence similarity, while strains of *A. sobria* and *A. veronii* biogroup *sobria*, which showed 60–65% DNA relatedness, were found to represent well-separated evolutionary lines, according to rDNA comparisons.

Pasteurellaceae

The three genera *Actinobacillus*, *Haemophilus* and *Pasteurella* were established between 1887 and 1917, quite early in the history of bacteriology, for isolates from veterinary and medical specimens and they were defined on the basis of origin, cultural characters, and microscopy. Emphasis on the sources of isolates in their classification led to such a massive proliferation in *Pasteurella* species that *Index Bergeyana* (1966, 1981) listed 117 mainly synonymous names! Furthermore, the agent of plague and its relatives (now in *Yersinia*) and the agent of tularaemia (now *Francisella tularensis*) were for some time allocated to *Pasteurella* because they were bipolar-staining, Gram-negative short rods, and the cause of whooping cough (now *Bordetella pertussis*) and its relatives and organisms now in *Moraxella* were initially accommodated in *Haemophilus*; these were removed to their new homes between the 1930s and 1960s.

The possibility of placing the three genera in one family was indicated by a numerical taxonomic study of a small number of species, and this suggestion was confirmed by DNA relatedness studies and further phenotypic characterization. All the strains are small, non-motile, fermentative rods or coccoid rods which are parasitic on the mucous membranes of humans and other animals and which show marked specificities for their host species. In 1944 André Lwoff suggested that in adapting to a parasitic way of life, members of these genera lost much of the genomic information carried by their free-living ancestors. Consequent limitations in their biosynthetic capabilities have exacerbated the problems of cultivation and phenotypic characterization of these fastidious organisms and have retarded progress in their classification and identification. Also, the isolation of new members of the family from a widening variety of hosts and niches is broadening the phenotypic span of the family and revealing species spectra and complexes.

By subjecting DNA reassociation binding-level data to single linkage cluster analysis, complex interrelationships were revealed at the 30% homology level and above. Six groups were indicated at the 40% homology level and it was suggested that they represented tribes; the tribe *Actinobacilleae* for example would contain the genus *Actinobacillus* with its species *A. lignieresii*, *A. equuli*, *A. suis* and *A. capsulatus*, whereas a new genus would be proposed for *A. actinomycetemcomitans* (since transferred to *Haemophilus*, however) and a *P. pneumotropica* biovar. Cluster analysis of homology data obtained under stringent conditions, using initial reassociation rates, revealed a distinct genus above the 50% level, *Pasteurella sensu stricto*, and allowed the recognition of 11 species and the division of *P. multocida* into three subspecies.

Such studies have shown that the genera and species are often no more distinct than those of the *Enterobacteriaceae*, but, with such a sturdy precedent and the value of nomenclatural stability in clinically important

Table 10.3 Changes within and between *Pasteurella*, *Actinobacillus* and *Haemophilus*, and clusters indicated by 16S rRNA sequence comparisons

Bergey's Manual (1974)	*Bergey's Manual* (1984)	Subsequent changes	rRNA cluster*
Pasteurella multocida	Five biovars	Three subspecies; biovar 1 → new genus?; biovar 6 →	3
		new species *P. canis*	3
P. haemolytica	Two (three?) biovars	Biovar A → *Actinobacillus* several new biovars → new genus with *P. testudinis*?	4
P. pneumotropica	→	→ *Actinobacillus*	5
P. ureae	→	→ *Actinobacillus*?	4
	P. aerogenes	→ new genus?	2
	P. gallinarum	→	3
		11 new species of which *P. avium* and *P. volantium* formed from *Haemophilus avium*; several new unnamed species	3
Actinobacillus lignieresii	→	→	4
A. equuli	→	→	4
"*A. suis*"†	*A. suis*	→	4
"*A. capsulatus*"	*A. capsulatus*	→	3
"*A. actinomycetemcomitans*"	"*A. actinomycetemcomitans*"	→ *Haemophilus*	1‡

continued on p. 156

Table 10.3 *continued*

Bergey's Manual (1974)	*Bergey's Manual* (1984)	Subsequent changes	rRNA cluster*
	"A. seminis"	*A. seminis* seven new species of which three transferred from *Pasteurella* and two from *Haemophilus*	2
Haemophilus influenzae	Six biovars	→	1
H. haemolyticus	→	→	1
H. parainfluenzae	Three biovars	→	4
H. haemoglobinophilus	→	→	3
H. aphrophilus	→	→	1‡
H. parahaemolyticus	→	→	4
H. paraphrophilus	→	→	1
H. paragallinarum	→	→ *Actinobacillus*	3
H. parasuis	→	→	3
H. paraphrohaemolyticus	→	→	4
H. gallinarum			
H. suis			
H. influenzae-murium			
H. ducreyi	→	→ New genus?	4‡
"H. aegyptius"	*H. aegyptius*	→	1
	H. segnis	→	1
	H. avium	→ *Pasteurella*	3
	H. pleuropneumoniae	→ *Actinobacillus*	4
	H. paracuniculus	→	3
	"H. equigenitalis"	→ New genus *Taylorella*	
	"H. agni"		
	"H. somnus"		2

*From Dewhirst *et al.* (1992); cluster 1, *Haemophilus sensu stricto*; 3, *Pasteurella sensu stricto*; 4, *Actinobacillus sensu stricto*.
† Quotation marks indicate species of uncertain affiliation.
‡ May warrant separate generic status on basis of DNA–rRNA hybridization; De Ley *et al.* (1990).

organisms, the three familiar generic names have been conserved in the family *Pasteurellaceae*. This stability is superficial, however, as numerous changes have occurred within the genera; not only have many new species and groups that probably represent new species been described, but there have also been several transfers or proposals for transfers of species between the three genera (Table 10.3).

The retention of the three genera, rather than including all in a single, redefined genus *Pasteurella*, brings practical problems since there are few, if any, phenotypic tests that correlate with the DNA relatedness groups. Numerical taxonomies, by and large, allow the biovars and species of *Pasteurella* and *Actinobacillus* to be distinguished, but not the genera. Furthermore, the traditional defining characters of *Haemophilus* species, the requirement for one or both of the growth factors V and X, have been called into question by the results of DNA relatedness studies, as organisms needing the factors are scattered among other members of the family and do not form a homogeneous group that is recognizable as a genus. For example, *H. pleuropneumoniae* requires V factor but was transferred to *Actinobacillus*, to which it showed phenotypic similarity, because it exhibited 90% DNA relatedness to a *P. haemolytica*-like organism and 80% relatedness to *A. lignieresii*, but only 30% relatedness to *H. influenzae*.

Similarly, *H. avium* and *H. paragallinarum* also require V factor but show higher relatedness to *Pasteurella* and *Actinobacillus*, respectively, than they do to *Haemophilus* species. Conversely, *A. actinomycetemcomitans* (there are few names longer) needs neither factor but has been transferred to *Haemophilus* on the basis of DNA relatedness, and common antigens, as demonstrated by immunodiffusion.

Chemotaxonomic approaches have met with limited success. Fatty acids help to define the family but not to separate its members, and cell sugars are more useful in this respect. The results of isoprenoid quinone analyses have emphasized the need for standardized methods. The absence of menaquinones, indicated by TLC and spectroscopy, was considered an important character of the family, separating it from the enterobacteria and some other groups, and organisms producing only menaquinones or ubiquinones and not demethylmenaquinones were excluded; the more sensitive HPLC method, however, showed menaquinones to be present in several members of the family as minor components, indicating that the picture is much more complicated than previously thought. It was, for example, initially suggested that *H. ducreyi* be removed to a new genus because it produced only menaquinones and ubiquinones; none the less, in this case such a move is also supported by DNA homologies of 0–6% with other members of the family.

Electrophoretic methods have been more helpful, with PAGE data supporting DNA relatedness used as evidence for transfers of species within the family, such as that of *H. pleuropneumoniae* to *Actinobacillus*, and assisting in the recognition of the V and X factor-independent species "*H. agni*" and "*H. somnus*" along with "*Histophilus ovis*" as representatives of a single species in a genus outside the family. Both these conclusions were supported by serological methods, such as immunoblotting in the later example. Using high resolution two-dimensional protein electrophoresis it has been possible to distinguish genera as well as species.

In genetic transformation experiments *Haemophilus* species (including *A. actinomycetemcomitans*), *P. multocida* and *P. pneumotropica* showed competition for homospecific transformation with *H. influenzae*, whereas

H. ducreyi, *H. parahaemolyticus*, *P. ureae* and *Actinobacillus* species showed little or no competition. This is consistent with the findings of DNA relatedness studies and suggests that *Actinobacillus* and *Haemophilus* are closer to *Pasteurella* than they are to each other.

Reorganization of the family along evolutionary lines may allow more phenotypic characters of differential value to be recognized. In the case of *H. aegyptius* it was suggested that it be regarded as a biovar of *H. influenzae* as the two were indistinguishable by DNA relatedness studies, but other workers believe that fresh isolates have sufficient distinctive properties to support the retention of species status. Phenotypic characters are still of great practical importance and multivariate analyses of enzyme tests and chemotaxonomic data show promise, while numerical taxonomies in conjunction with nucleic acid studies have been valuable in confirming some old species of uncertain affiliation and revealing some new taxa. Thus, the original strain of "*A. seminis*" clearly does belong to *Actinobacillus* although many other strains allocated to the species do not, and "*H. equigenitalis*" has been placed in the new genus *Taylorella*.

Phylogenetic information from 16S and 23S rRNA hybridizations with DNA, and 16S rRNA sequencing confirm the validity of the family *Pasteurellaceae*, but distinction between the genera is obscured by the intermediate positions of many species between genera and the existence of three main branches containing organisms currently allocated to the genus *Haemophilus*. However, sequencing work has supported the transfers of *P. ureae* to *Actinobacillus* and *A. actinomycetemcomitans* to *Haemophilus* and subsequently allowed the recognition of five clusters in the family (Table 10.3). Clusters 1, 3 and 4 represent *Haemophilus*, *Pasteurella* and *Actinobacillus* respectively, and it is clear that many nomenclatural changes are needed. These can follow once the many unnamed taxa have been rRNA sequenced and the genomic species matched with phenotypic characters. The resolution of the phylogenetic, and so taxonomic, picture is not just an academic exercise as several of the organisms whose affiliations are uncertain and which await valid publication are of veterinary or medical importance; for example, "*H. agni*", "*H. somnus*" and "*Histophilus ovis*", cause ovine septicaemia, bovine infectious thromboembolic meningoencephalitis and various ovine infections, respectively, and yet are considered to represent one taxon.

Bacteroidaceae

The earth's atmosphere was anaerobic for the larger part of its history, resulting in the evolution of a great diversity of anaerobic bacteria. Difficulties of studying such organisms, because of the problems of anaerobic cultivation, have retarded systematic work. For the Gram-negative rods, be they straight, curved or helical, round-ended or fusiform, classification schemes at the genus level were based on a few fairly easily determined characters such as shape, flagellation, or fermentation products, and yet great variations in size and shape occurred within genera. In *Bergey's Manual* (1984) 13 genera were listed in *Bacteroidaceae*, but some doubts were entertained about the inclusion of *Acetivibrio*, *Butyrivibrio* and *Lachnospira* species because their cell wall structures resemble those of Gram-positive bacteria.

In recent years technical advances in the cultivation and manipulation of anaerobes, and a concomitant increase in awareness of their medical significance, have encouraged the systematic application of computer taxonomy, and chemotaxonomic and genotypic methods. The inadequacies of the previous classification were readily revealed and now the definitions of the Gram-negative anaerobic genera are becoming precise and well-informed. Work has concentrated on the two genera of most clinical significance, *Bacteroides* and *Fusobacterium*.

Of the 44 species listed in *Bergey's Manual* (1984), 26 have been reclassified in 11 new genera or found to be synonymous and so combined in one species, but this has been counteracted by the description of 18 new species. Furthermore, several new genera based upon new isolates have been proposed so that the numbers of species in *Bacteroides* and genera in *Bacteroidaceae* continue to increase. The family's boundaries cannot be determined until the genera have been clearly defined and their relatedness established, but great strides in this direction have already been made.

It has been proposed that the genus *Bacteroides* should be restricted to *B. fragilis* and its present nine close relatives, so that the G+C range for the genus would shrink from 28–61 mol% to 39–48 mol%. The genus would then contain saccharolytic species, yielding acetic and succinic acids as major fermentation products, that possess pentose phosphate pathway enzymes such as glucose-6-phosphate dehydrogenase (G6PDH) and 6-phosphogluconate dehydrogenase (6PGDH) (absent from all other species studied except *B. hypermegas*, whose enzyme species are different), sphingolipids, predominantly methyl branched long-chain fatty acids, and menaquinones as their sole respiratory quinones. This proposal is supported by DNA–rRNA hybridization, 16S rRNA cataloguing and 5S rRNA sequencing of a limited number of strains which show the *B. fragilis* group to represent a distinct line branching deeply in superfamily V (the *Cytophaga–Flavobacterium* phylum).

Other rRNA studies have led not only to the reclassification of species in new genera (Table 10.4), but have also shown that some species belong to quite different evolutionary lines. *Bacteroides amylophilus* has predominantly straight-chain fatty acids, does not possess sphingolipids, and belongs to the γ subclass of the *Proteobacteria* (superfamily I); it is more closely related to *E. coli* than it is to *B. fragilis* and has been reclassified in the new genus *Ruminobacter* (Fig. 10.1). Partial sequence analysis of a rRNA gene of *B. nodosus* (now *Dichelobacter nodosus*) showed that it too belongs in the γ subclass and sequencing of 16S rRNA revealed that "*B. gracilis*" and "*B. ureolyticus*" are closely related to the campylobacters.

Other new genera have been proposed chiefly on the basis of biochemical and chemotaxonomic characters, supported by data from rRNA studies as they become available. One is *Prevotella*, which contains the 16present members of the *B. melaninogenicus–B. oralis* group, only some of which produce pigmented colonies on blood agar and all of which are

Table 10.4 Changes in the classification of *Bacteroides* species

Bacteroides species	Source or habitat	Taxonomic status	Year of proposal
B. hypermegas	Human and animal gut	New genus *Megamonas*	1982
B. multiacidus	Human and pig faeces	New genus *Mitsuokella*	1982
B. microfusus	Animal faeces	New genus *Rikenella*	1985
B. furcosus	Clinical specimens, faeces and snails	New genus *Anaerorhabdus*	1986
B. amylophilus	Rumen	New genus *Ruminobacter*	1986
B. termitidis	Termite gut	New genus *Sebaldella*	1986
B. praeacutus	Clinical specimens	New genus *Tissierella*	1986
B. asaccharolyticus, *B. endodontalis* and *B. gingivalis*	Human oral cavity and clinical specimens	New genus *Porphyromonas*	1988
B. succinogenes	Rumen	New genus *Fibrobacter*	1988
B. fragilis and nine other species	Human faeces and clinical specimens	*Bacteroides* (emended definition)	1989
B. melaninogenicus, *B. oralis* and 14 other species	Human gingival crevice and clinical specimens, and rumen	New genus *Prevotella*	1990
B. nodosus	Bovine, caprine and ovine footrot	New genus *Dichelobacter*	1990
B. levii, *B. macacae* and *B. salivosus*	Animals, oral cavity and rumen	Related to *Porphyromonas*?	
"*B. gracilis*" and "*B. ureolyticus*"	Human oral cavity and clinical specimens	Related to *Campylobacter*, not anaerobic but microaerophilic	
>15 other species	Animals and human clinical specimens	Not *Bacteroides*; affiliation unknown	

phenotypically similar, being bile-sensitive, moderately saccharolytic, and lacking the enzymes G6PDH and 6PGDH. *Prevotella* lies in the same part of superfamily V as *Bacteroides sensu stricto* and so may be a member of an emended *Bacteroidaceae*. *Porphyromonas* was proposed for the asaccharolytic or weakly saccharolytic, pigmented species *B. asaccharolyticus*, *B. endodontalis* and *B. gingivalis*. This still leaves at least 15 species that do not belong to *Bacteroides*, or any of the newly proposed genera, and whose generic positions are unknown, and yet new species continue to be assigned to the genus even though several bear little similarity to *B. fragilis*. In addition, new genera of Gram-negative anaerobes are regularly described, and their affiliations need to be determined.

Fusobacterium has been studied less intensively, but with the removal of three spore-bearing species to *Clostridium*, its taxonomy is more satisfactory than that of *Bacteroides*. There are 12 species, with a range of 26–34 mol% G+C for 11 of them and, although cell wall composition and glutamate dissimilation mechanisms indicate heterogeneity, rRNA sequence comparisons are consistent with a single genus, albeit one containing at least five groups of species. A 16S rRNA cataloguing study indicated that *F. nucleatum* (the type species) belongs to the *Bacteroides–Cytophaga–Flavobacterium* evolutionary line (superfamily V) in a subgroup that contains representatives of the genus *Prevotella*, but comparison of 5S rRNA sequences places *Fusobacterium* species in a subgroup quite separate from the main *Bacteroides–Porphyromonas– Cytophaga* cluster.

Important species

From the foregoing it will be appreciated that many of the organisms considered in this chapter have diverged comparatively recently in evolutionary time so as to exploit different environmental niches in animals and humans. Some are successful as commensals whereas others are characteristically pathogenic; the difference in lifestyle may be at strain rather than species level and may be dependent more upon circumstances than invasiveness and virulence. Several species were major killers in earlier times and remain important agents of disease in those countries that lack clean water supplies and have poor sanitation.

Members of *Enterobacteriaceae* are chiefly important as causes of diarrhoeal illness. *Escherichia coli* comprises about 1% of the intestinal flora of warm blooded animals, and until the 1940s it was regarded as a harmless gut commensal compared with the two major pathogens *Salmonella* and *Shigella*. It is now appreciated that *E. coli* is an important and wide-ranging pathogen so that the distinction between the genera is less clear. Enterotoxigenic, enteropathogenic and enteroinvasive strains of *E. coli*, (properties associated with particular antigenic types) are major causes of gastroenteritis; the first of these groups, by dehydration and shock, kills millions of children each year in some economically disadvantaged countries, and it is the most common cause of traveller's diarrhoea. Similar organisms cause, for example, systemic or enteric colibacillosis in domestic animals and white scour of calves. Verotoxin-producing strains cause haemorrhagic colitis which may be followed by haemolytic uraemic syndrome, an important cause of acute renal failure. Other strains cause a variety of opportunistic infections including human cystitis (where it is pre-eminent) and neonatal meningitis, cystitis and pyometritis of dogs and cats, joint ill of foals, and bovine and porcine mastitis. *Salmonella* is a household word synonymous, for most people, with food-borne infection, and not without reason, but few realize how serious the illness can be, with death from septicaemia and other complications, or that domestic animals are not just reservoirs of infection but also suffer from enteritis and other problems such as abortion. *Klebsiella pneumoniae* is another wide-ranging pathogen in humans and animals causing, in particular, respiratory infections. Other enterobacterial species such as *Proteus* and *Serratia* are also important opportunistic pathogens, as Table 10.1 indicates. *Salmonella typhi* is a human pathogen only, and typhoid and the paratyphoids (the enteric fevers) remain serious endemic diseases in many parts of the world. Bacillary dysentery, caused by *Shigella* species, occurs throughout the world and the more serious forms caused by *S. dysenteriae* serovar 1 can result in high mortality rates if untreated.

Yersinia species seem to represent a spectrum of adaptation to a parasitic mode of existence. *Yersinia pestis* infection is endemic in many wild rodent populations and occasional epizootics can spread to domestic rodents (rats) and so to humans by flea bite or direct contact. Bubonic and pneumonic plague are chiefly of historical importance, as they are believed to be largely responsible for the black death pandemics of previous centuries, but outbreaks still occur in some parts of the world. *Yersinia pseudotuberculosis* and *Y. enterocolitica* are found in the intestines of a wide variety of animals in which they may cause adenitis and diarrhoea, and oral transmission to humans can occur. Increased awareness has revealed that human infections are not uncommon; mesenteric adenitis simulating appendicitis and acute gastroenteritis are the usual symptoms, but infections may progress to arthritis and septicaemia in compromised persons.

Erwinia species are parasites and pathogens of plants rather than animals. They cause considerable economic loss through blights, cankers, die backs, wilts, soft rots, discolourations and leaf spots, such as fireblight of pome fruit trees (*E. amylovora*), vascular wilt of willows (*E. salicis*) and rotting of potatoes, carrots, etc. (*E. carotovora*).

Members of the *Vibrionaceae* are common in marine and freshwater environments and mainly behave as saprophytes, but toxigenic serovar O:1 strains of *V. cholerae* cause cholera in humans. Two biovars are recognized but taxonomic studies do not support their separation; they are the Classical biovar, which caused pandemics and epidemics during the 18th and 19th centuries, and the Eltor biovar which is endemic in India and South-east Asia and responsible for the present pandemic and an epidemic in South America. Other serovars can cause cholera-like diarrhoea. *Vibrio parahaemolyticus* causes food-borne infection from seafood and other species can cause opportunistic infections. Various species have been found in diseased sea creatures and *V. anguillarum* is an important cause of acute septicaemia in free-living and farmed marine fish and eels. Strains of several *Vibrio* species, *Photobacterium* species and *Xenorhabdus luminescens* are bioluminescent, the *Photobacterium* species being symbionts in the luminous organs of certain marine fish. Bioluminescence is exploited in a wide range of techniques for enumerating bacteria and measuring chemical agents or pollutants. Aeromonads are found in the freshwater environment; *A. salmonicida* subspecies are strict parasites of fish and are particularly important as the agents of furunculosis in salmonid fishes and as pathogens of other fish, and *A. hydrophila* is pathogenic for fish and frogs and occasionally infects humans and other mammals.

Pasteurellas are widely distributed as commensals of animals but may often behave as secondary invaders or opportunists, and several species are primary pathogens of domestic animals. Different serovars of *P. multocida* are responsible for haemorrhagic septicaemia of cattle and fowl cholera.

The organism is also a secondary invader in the respiratory system of stressed animals where it causes severe pneumonia, an opportunist in many animals including humans, and a frequent contaminant of dog and cat bites. *Pasteurella haemolytica* is particularly important as an agent of shipping fever, a pneumonia of cattle, sheep and goats, and it can also infect other animals. Haemophilas likewise form part of the indigenous flora of the mouths and upper respiratory tracts of many animals and may be opportunistic. *Haemophilus parasuis*, for example, is the agent of Glässer's disease which is a systemic fibrinous inflammatory condition of stressed pigs. Of the several species carried by humans, *H. influenzae* is the most important as capsulate strains are primary pathogens causing meningitis and other septicaemic conditions, particularly in children. Non-capsulate strains may be secondary invaders and are often implicated in sinusitis, bronchitis, conjunctivitis, otitis media and other infections. *Haemophilus aegyptius* is a close relative of *H. influenzae* and causes conjunctivitis. *Haemophilus ducreyi* causes the venereal disease chancroid and several other species are occasional opportunists.

Actinobacillus species are also commensals in the mouths, guts, upper respiratory and female genital tracts of animals, and several are opportunistically pathogenic. *Actinobacillus lignieresii* is the best known, causing chronic granulomatous lesions in cattle (wooden tongue), sheep and, occasionally, in other animals and humans. The species *A. pleuropneumoniae* (formerly in *Haemophilus*) is more of a primary pathogen, as it is responsible for serious and very contagious pneumonia and pleuritis in unstressed pigs. Another respiratory tract infection is fowl coryza, a catarrhal condition caused by *A. paragallinarum. Actinobacillus equuli* and *A. suis* are associated with septicaemic conditions of horses (purulent nephritis, arthritis and joint-ill of foals, and endocarditis, meningitis and metritis of adult animals) and pigs (pneumonia and arthritis), respectively, but each can cause similar conditions in the other animal as well.

Gram-negative anaerobic rods are the most frequently isolated anaerobes in clinical infections and members of *Bacteroides sensu stricto* are members of the intestinal flora and the types most commonly recovered from specimens, particularly those from intra-abdominal infections. Members of the bile-sensitive genera form part of the oral, dental and genitourinary tract floras, and some species, particularly *Prevotella* species, *Fusobacterium nucleatum* and *Porphyromonas gingivalis*, are important opportunistic pathogens of their habitats. Species such as *Bacteroides ruminicola*, *Fibrobacter succinogenes* and *Ruminobacter amylophilus* are found in the rumen where they play an important part in the degradation of carbohydrates and protein, whereas *Dichelobacter nodosus* is a major pathogen, causing footrot in cattle, goats and sheep.

Identification

With so many of the organisms considered in this chapter being of medical and veterinary importance it is to be expected that much effort has been expended on the development of rapid and simple identification schemes for routine use. Indeed, the commercial development of identification kits of all kinds initially concentrated on systems for the enterobacteria and other Gram-negative, fermentative rods, and such kits still account for the largest part of the identification kit market.

Identification to species level, and beyond that by serotyping, biotyping, bacteriophage typing or bacteriocin typing for epidemiological purposes, is essential for most infections with these organisms, and the results are needed urgently if they are to be of value. Thus step-by-step approaches requiring a week or more have been replaced by unitary procedures that give results within 24 hours of pure culture. Suspected *Enterobacteriaceae* members can be checked on isolation plates with a spot oxidase test, and *E. coli* and swarming *Proteus* species, which are some of the most common clinical isolates of this family, are readily recognized by colony morphology and a spot indole test, so that further, time-consuming testing is unnecessary.

Most clinical laboratories use commercial kits that are usually based upon conventional biochemical tests such as carbohydrate fermentations, hydrolyses, H_2S production, and indole; by using heavy inocula (and special substrates sometimes), or by adopting a different approach and detecting preformed enzymes, results may be obtained within 4 hours and identities furnished by regularly updated computer databases. Some manufacturers have automated these methods, the reading and interpretation being done by machine. Traditional methods using tubed media and combination media are still widely used, however, because they are cheap, and for halophilic vibrios, which require some salt in the medium, they can be more reliable.

Agglutination tests may then be used to type *Salmonella* and *Shigella* species, for example, with immediate results, but enterotoxigenic *E. coli* are more difficult to recognize and may require ELISA or DNA probes. The other typing methods require further overnight incubations and a drawback with all the methods mentioned is the need for a pure culture; consequently DNA probes, assisted by PCR amplification, and other approaches which obviate a cultivation step are being developed for some of the more important pathogens.

The clinically important strict anaerobes can be readily identified to genus by their microscopic and colonial morphologies, sensitivities in antibiotic disk tests and a small number of biochemical tests; definitive identification requires further biochemical tests and analysis of metabolic end products by GLC, but whole-organism fatty acid and sugar patterns, enzyme electrophoresis, PMS, restriction endonuclease analysis, and chromosomal DNA probing by dot blot assay have all shown potential. Several commercial kits of biochemical tests are available, those based upon preformed enzymes rather than fermentation tests being faster and more useful for identifying asaccharolytic species.

11 The Gram-positive rods

Whether we will see in the future a genetic relationship between the hay bacillus and the anthrax bacillus . . . or whether we are dealing here with outwardly similar but specifically different species or races, will have to be left to the further development of science to decide.

Ferdinand Cohn, 1876, *Studies on the Biology of the Bacilli*

Introduction

Cell shape and sporulation have been emphasized in the taxonomy of the Gram-positive rods, but if these features are ignored resemblances between certain species in apparently quite distinct genera, such as *Bacillus*, *Sporolactobacillus* and *Lactobacillus*, or *Bacillus*, *Sporosarcina* and *Planococcus*, become very strong. For example, sporulating lactic acid bacteria isolated in the 1960s raised doubts about the boundaries of *Bacillus* (defined as catalase positive, usually motile aerobes and facultative anaerobes) and *Lactobacillus* (defined as catalase negative and rarely motile microaerophiles). A spectrum of such organisms ranging from typical *Bacillus* species to typical *Lactobacillus* species could be recognized (Table 11.1), and *Sporolactobacillus* appeared closer to *Bacillus*; however, an early oligonucleotide cataloguing study indicated that it was only distantly related.

As 16S rRNA sequencing comparisons progressed it became clear that the familiar endospore-formers, and the lactobacilli, streptococci and

Table 11.1 Some characters of *Bacillus–Lactobacillus* intermediates

	Lactate production	Catalase production	Spore formation	Fatty acid pattern*	Menaquinones present	Motility	DAP in cell wall
Typical *Bacillus*	–	+	+	B	+	+	+
B. coagulans	+	+	+	B	+	+	+
B. racemilacticus	+	+	+	B	+	+	+
Sporolactobacillus inulinus	+	–	+	B	+	+	+
Lactobacillus yamanashiensis	+	–	–	L	+	+	+
L. plantarum	+	–	–	L	–	–	+
Typical *Lactobacillus*	+	–	–	L	–	–	–

* *Bacillus* pattern is predominantly saturated fatty acids with odd numbers of carbon atoms, and with iso- and anteiso-branched C_{15} and C_{17} and smaller amounts of iso-C_{16}; *Lactobacillus* pattern is even-numbered, saturated and unsaturated (mainly C_{16}, and C_{16} and C_{18} respectively) with C_{17} and C_{19} cyclopropane acids.

staphylococci all belong to the same low mol% G+C subdivision of the Gram-positive bacteria, and they appeared to represent well-defined if sometimes rather loose lineages. However, in several studies non-sporeformers were found to be intermingled with sporeformers, and when heterogeneous taxa such as *Bacillus* were represented by a wide range of their component species, very complex pictures began to emerge, indicating that many new genera will ultimately have to be recognized.

Other well-established divisions have also been shown to be unnatural, emphasizing again that morphology is not a good indicator of suprageneric relationships. A glance at Fig. 11.1 indicates that placement of rods in this chapter and cocci in the next is an arbitrary separation that does not accurately reflect phylogenetic relationships.

The endospore-formers

The only character that members of this very diverse assemblage have in common is the ability to produce endospores. Although most are Gram-positive motile rods, loss of positivity, Gram-variability and Gram-negativity are not uncommon, some of the most important species are non-motile, and morphologies range from cocci to filaments; thus the group is something of a dumping-ground for endospore-formers. Its subdivision on the basis of physiology and cell shape, into aerobic and facultatively anaerobic rods (*Bacillus*), microaerophilic rods (*Sporolactobacillus*), anaerobic rods that do not reduce sulphur to sulphide (*Clostridium*) and those that do (*Desulfotomaculum*), and aerobic cocci (*Sporosarcina*) is, of course, useful for identification. Spore formation, likewise, remains a valuable aid to identification, but certain clostridia in particular sporulate only feebly, and it is quite likely that established non-sporulating genera harbour asporogenous clostridia and perhaps bacilli. Indeed, spores have been reported in *Sarcina*, a genus of anaerobic cocci that is phylogenetically close to *Clostridium butyricum*. Meantime, other related but asporogenous genera such as *Acetobacterium*, *Eubacterium*, *Planococcus* and *Ruminococcus* are excluded from the group. Also excluded, because it was classified with the actinomycetes, is the endospore-former *Thermoactinomyces*.

The internal classification of the group is most unsatisfactory: not only do the main genera show wide ranges of mol% G+C, 32–69 for *Bacillus* and 22–55 for *Clostridium*, but so do some of the 'species'; for example, *Bacillus circulans* shows a range of 31–61 mol% G+C, and has been separated into genomic species. Furthermore, there have been several reports of what appear to be endospore-formers of various morphologies from animal guts; they are often very large and some contain several spores, but they have yet to be cultivated and their affiliations are therefore unknown. They include *Oscillospira*, whose fat rods are divided into disk-shaped cells, "*Anisomitus*", "*Arthromitus*" and "*Coleomitus*" which also produce septate filaments or trichomes, "*Fusosporus*" and "*Sporospirillum*", which have large, fusiform or helical cells containing one or two spores, and "*Metabacterium*" whose large rods contain up to eight spores. It is perhaps worth

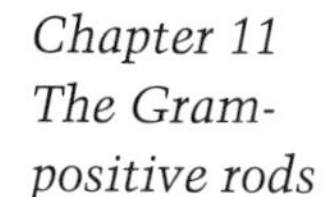

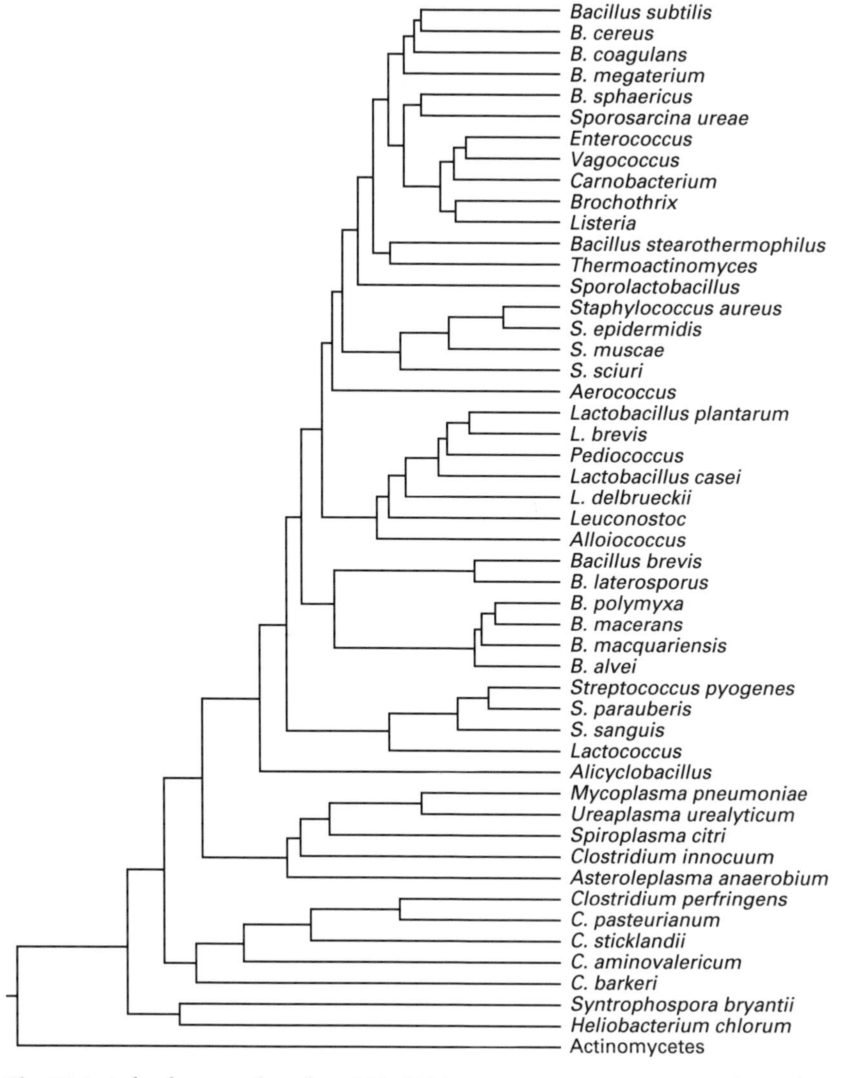

Fig. 11.1 A dendrogram, based on 16S rRNA sequence comparisons, to show the probable evolutionary relationships between the main lineages of the Gram-positive bacteria and mollicutes. In several instances the branching orders at lower levels of relatedness are somewhat speculative.

noting that two trichome-forming, asporogenous organisms have been shown to be phylogenetically related to *Bacillus* species (see *Genomic Analyses* below).

There have been many taxonomic studies of endospore-formers, *Bacillus* and *Clostridium* especially, but few were comprehensive and it is generally agreed, even by ardent lumpers, that these two genera require splitting. That no successful attempts have yet been made to do this reflects the very great diversity of the organisms, and the necessary studies based upon genomic analyses such as DNA–DNA and DNA–rRNA homologies, rDNA restriction endonuclease analysis, and rRNA sequencing of carefully selected and representative strains, are enormous tasks only recently begun.

Bacillus and relatives

As well as having a mol% G+C range that spans at least four genera (on the basis of the theoretical limit of 10% for a genus), *Bacillus* encompasses a wide variety of physiological types including psychrophiles, mesophiles and thermophiles, acidophiles and alkaliphiles, strict aerobes and facultative anaerobes, with nutritional requirements which range from facultative chemolithotrophy through versatile heterotrophy and fastidious heterotrophy to obligate parasitism. Given this breadth, the exclusion of *Sporosarcina* on grounds of shape, and of other genera because they are non-sporeforming, seems unreasonable. However, any reallocation of such taxa must await the outcome of the major task of dividing *Bacillus* itself, which contains over 65 validly published species and over 30 others that await full study. Although most taxonomic studies support splitting of the genus, the boundaries are unclear and more genomic studies are needed; none the less, some basic divisions are apparent.

Phenotypic Analyses. The few numerical taxonomies that have approached comprehensiveness are largely congruent and indicate the existence of six or more groups of species which correlate with those identified primarily on the basis of microscopic morphology in the pioneering studies of Gibson, Smith and Gordon between the 1930s and 1970s:

1 The *B. cereus* group, containing its close relatives *B. anthracis*, *B. thuringiensis* (whose pathogenicities are plasmid-mediated) and *B. mycoides* which some regard as subspecies of *B. cereus*; large-celled species with ellipsoidal spores that do not swell the sporangium (Fig. 11.2).

2 The *B. subtilis* group, including *B. amyloliquefaciens*, *B. licheniformis* and *B. pumilus* which have small cells with ellipsoidal or cylindrical spores that do not swell the sporangium (Fig. 11.2); *B. megaterium*, although it is morphologically more similar to *B. cereus*, is affiliated to this group, and groups 1 and 2 may be closely related.

3 The *B. circulans* group includes *B. macerans* and *B. polymyxa*; their ellipsoidal spores swell the sporangia (Fig. 11.2).

4 *Bacillus coagulans* and *B. stearothermophilus* are thermophilic and produce ellipsoidal spores that swell the sporangia.

5 & 6 The compositions of the remaining two groups vary between studies so that the affiliations of several species are unclear, but one group (5) contains *B. sphaericus* and other species whose spherical spores swell the sporangia (Fig. 11.2), and the other includes *B. firmus* whose morphology is similar to group (2) species. Distinction within and between groups may be somewhat obscured owing to the heterogeneity of some species, and several of these so-called spectra or complexes have been subdivided on the basis of DNA relatedness.

Chemotaxonomic studies of the genus have been limited, and the available data, much of them generated by non-taxonomic investigations, have been obtained by a variety of methods. Results are therefore difficult to interpret with heterogeneous species, and of variable value overall. A

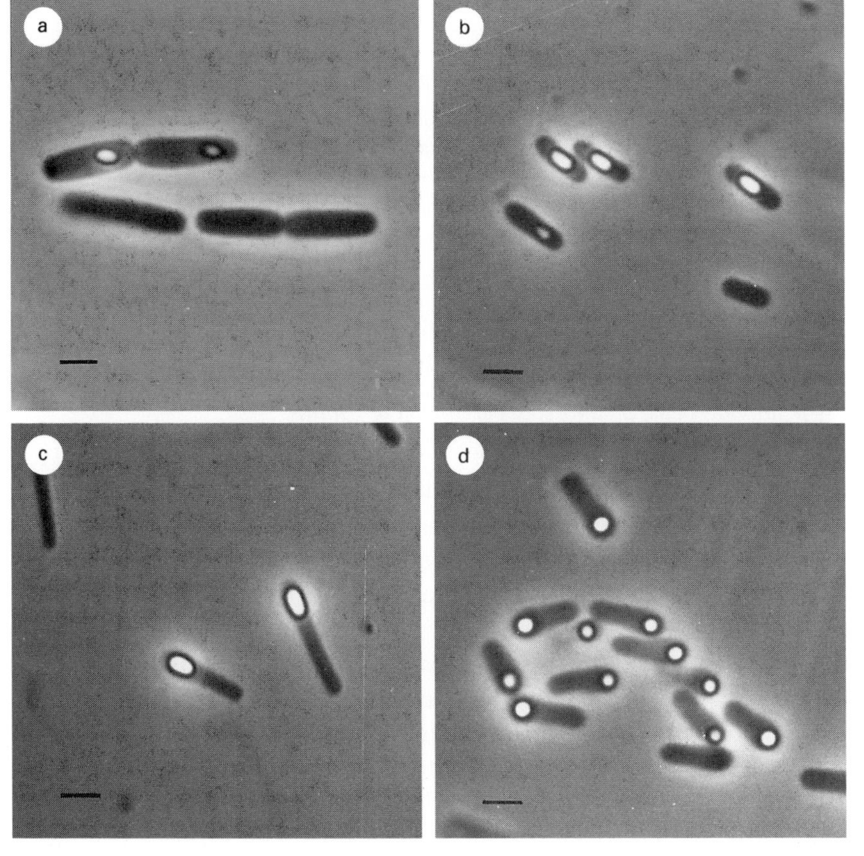

Fig. 11.2 Sporangial morphologies of the type strains of a, *Bacillus cereus*; b, *B. subtilis*; c, *B. circulans*; and d, *B. sphaericus*. Bar markers represent 2.0 μm.

further problem with such a varied group is the impossibility of standardizing growth conditions. Peptidoglycan and major menaquinone types are broadly similar across the species studied and, along with fatty acid patterns, imply that *Sporolactobacillus* is closer to *Bacillus* than it is to *Lactobacillus*. Atypical peptidoglycans are found in certain round-spored species, the cross-linkages containing lysine or ornithine rather than *meso*-DAP. Fatty acid, polar lipid, minor menaquinone, and enzyme profiles show more variation and have been of taxonomic value within some groups and species; whole-organism fatty acid patterns, for example, can separate the close relatives *B. cereus* and *B. anthracis*.

Genomic Analyses. The heterogeneity of the genus indicated by the mol% G+C range is confirmed by DNA relatedness studies. There have been no systematic relatedness studies of the homogeneous species, but with some of the heterogeneous species this approach has enabled the recognition of groupings that correlate with phenotypic characters and may warrant species status (Table 11.2); thus, *B. amyloliquefaciens* has been established as a species separate from *B. subtilis*, a *B. coagulans* subgroup forms the new species *B. smithii*, and *B. circulans* represented at least 10 species.

Table 11.2 DNA relatedness groups of some *Bacillus* species

Species	Mol% G+C range: Species	Mol% G+C range: Strains in relatedness study	Number of strains	Relatedness groups: Groupable strains	Relatedness groups: Groups of >1
B. subtilis	41–48	41–45	27	26	2
B. megaterium	34–48	34–41	20	19	3
B. coagulans	37–56	37–47	82	82	3*
B. sphaericus	34–40	34–37	62	49	6
B. circulans	31–61	37–55	111	50	10†

* One group forms the species *B. smithii*.
† Four groups reclassified as *B. amylolyticus*, *B. lautus*, *B. pabuli* and *B. validus*.

Until recently there has been surprisingly little phylogenetic study of the genus, no doubt owing in part to its unmanageable size; 16S rRNA oligonucleotide catalogues were prepared for only one or two representatives from less than a dozen species, five of which are heterogeneous. These few data suggested that the *B. cereus* and *B. subtilis* groups are related, though more distantly than genera in the *Enterobacteriaceae*. Conversely, *B. pasteurii* appeared closer to *S. ureae*, *B. sphaericus* closer to *Caryophanon latum* (asporogenous, multicellular rods or trichomes), *B. globisporus* closer to *Filibacter limicola* (asporogenous, gliding trichomes), and "*B. aminovorans*", and to a lesser extent *B. insolitus*, closer to *Planococcus citreus* (asporogenous, marine organism), than they do to other *Bacillus* species. It is interesting that peptidoglycan types correlate with most of these groupings, but perplexing that this is not the case with the last of them, which also shows a wide range of mol% G+C. Further distant are *B. stearothermophilus*, which showed some relatedness to *Thermoactinomyces vulgaris*, and the two rapidly evolving species *B. coagulans* and *Sporolactobacillus inulinus*.

These findings have mostly been supported by comparisons of 16S rRNA sequence analyses of the type strains of a wider selection of species, which have indicated several groupings that are in agreement with those based on phenotypic characters (Fig. 11.3). Both approaches recognize a *B. cereus* group and a *B. subtilis* group, and a comparative sequencing study showed that these two groups are indeed quite closely related, but are distant from *B. megaterium*, and all belong to a large cluster comprising representatives of 27 species. Many of these species seemed to diverge from the same area of the tree, so that subclusters could not be discerned and new genera might be difficult to define.

Another group contained *B. sphaericus* and other round-spored species including *B. pasteurii* and *Sporosarcina*, also predicted by phenotypic analyses, while *B. macerans*, *B. polymyxa* and the two species *B. amylolyticus* and *B. pabuli* (which were derived from the *B. circulans* complex) are also phylogenetically close as expected. The *B. circulans* strain was found in a different cluster (and, therefore, genus), emphasizing the uncertainty that may result from working with a single representative of a taxon.

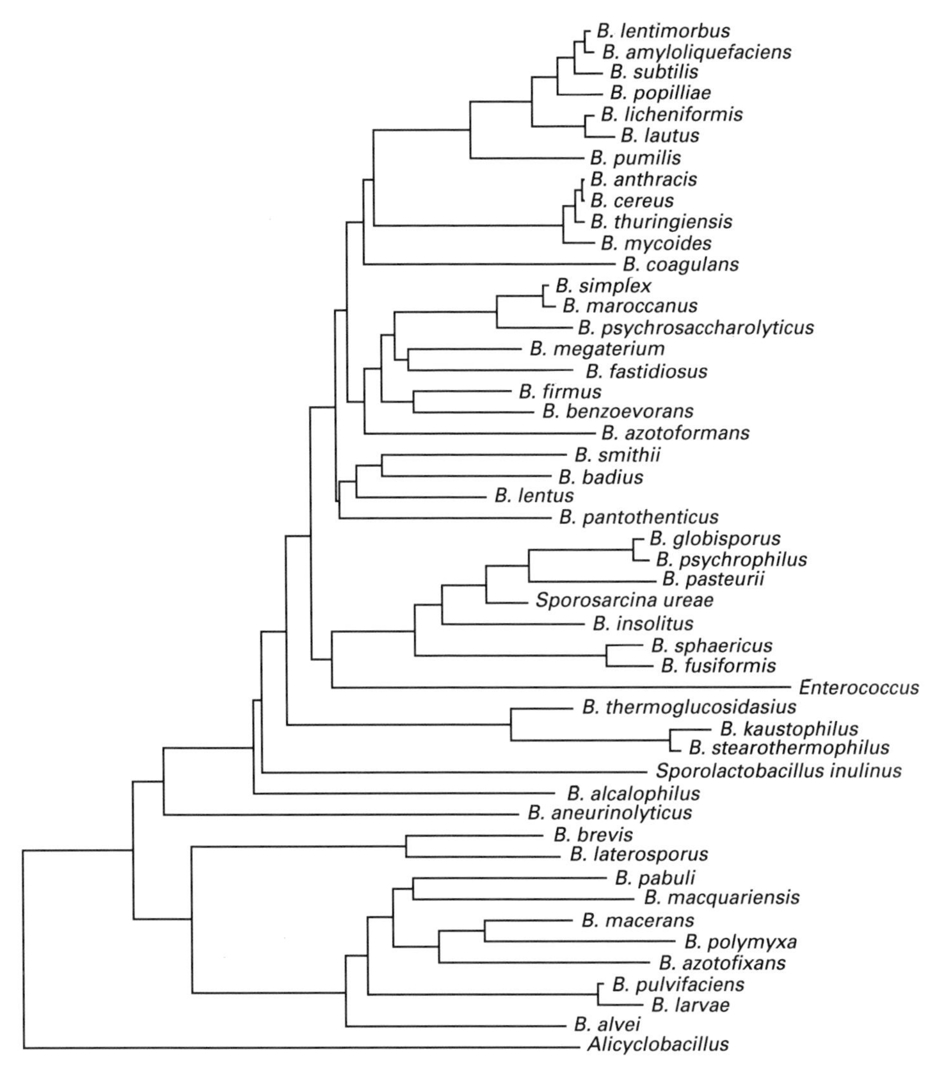

Fig. 11.3 Evolutionary tree, based on 16S rRNA sequence comparisons, for *Bacillus* species and some related genera. Adapted from Ash *et al.* (1991).

The thermophiles *B. coagulans* and *B. stearothermophilus* however, are distant from each other. Another thermophile, *B. cycloheptanicus*, represented a distinct and rapidly evolving line, and a separate sequencing study found it to be related to two other thermophilic species, *B. acidocaldarius* and *B. acidoterrestris*. These three species have ω-alicyclic fatty acid as the major membrane lipid component and the new genus *Alicyclobacillus* has been proposed for them.

It seems that when *Bacillus* is eventually split into the several genera it undoubtedly represents, a phylogenetic approach might sometimes prove difficult to combine with a phenotypic one that emphasizes easily-determined characters such as sporulation because of their diagnostic usefulness. Genera will have to be separated on the basis of a pattern of properties rather than single, unique characters. Compromises will be required, therefore, as practical considerations dictate that genera and species be phenotypically coherent. Thus, a tentative arrangement has been suggested in which a redefined *Bacillaceae* contains *Bacillus* (and whatever new genera it may yield), *Sporosarcina*, *Sporolactobacillus*, *Thermoactinomyces* and such non-sporeformers as *Listeria*, *Brochothrix* and *Staphylococcus* (Table 11.3).

It has also been confirmed that 16S rRNA sequence comparisons, and sometimes DNA relatedness data, may fail to separate closely related (i.e.

Table 11.3 An illustration of how the clostridia and relatives might be classified according to phylogenetic and phenotypic characters

Order	Family	Examples of genera and species
1 *Clostridiales*	1 *Clostridiaceae* I	*Clostridium butyricum, C. perfringens* (group I), *Sarcina*
	2 *Clostridiaceae* II	*C. difficile, C. lituseburense* (group II-A), *Eubacterium tenue, Peptostreptococcus anaerobius*
	3 *Clostridiaceae* III	*C. oroticum, C. sphenoides* (group III), *Peptococcus glycinophilus, Streptococcus hansenii*
2 *Desulfotomaculales*	1 *Desulfotomaculaceae*	*Desulfotomaculum, Ruminococcus, (Peptococcus aerogenes?)*
3 *Mycoplasmatales*	1 *Mycoplasmataceae*	*Acholeplasma, Anaeroplasma, Mycoplasma, Spiroplasma, Ureaplasma*
	2 *Clostridiaceae* IV	*C. innocuum, C. ramosum* (group VI), *Erysipelothrix, Lactobacillus catenaforme*
4 *Acetobacteriales*	1 *Acetobacteriaceae*	*Acetobacterium, C. barkeri, Eubacterium* (group IV)
	2 *Thermoanaerobiaceae*	*Acetogenium, C. thermoaceticum* (group V), *Thermoanaerobium*
5 *Bacillales*	1 *Bacillaceae*	*Bacillus, Brochothrix, Caryophanon, Filibacter, Gemella, Listeria, Planococcus, Sporolactobacillus, Sporosarcina, Staphylococcus, Thermoactinomyces*
	2 *Streptococcaceae*	*Enterococcus, Lactococcus, Streptococcus*
	3 *Lactobacillaceae*	*Lactobacillus, Leuconostoc, Pediococcus*
	Uncertain affiliation	*Aerococcus, Kurthia*
Uncertain affiliation: *Butyrivibrio, Heliobacillus, Heliobacterium, Selenomonas, Sporomusa*		

Adapted from Cato & Stackebrandt (1989).

recently diverged) species that are separable by phenotypic characters. Whether or not *B. anthracis*, *B. cereus*, *B. mycoides* and *B. thuringiensis* should be recognized as separate species (as is practically sensible) has been disputed for many years, and the reduction of *B. anthracis* to subspecies was proposed on the basis of DNA relatedness; 16S rRNA sequences of a *B. cereus* from emetic food poisoning and a *B. anthracis* (whose pathogenicity is plasmid-borne) were found to be identical over a stretch of 1446 bases and to differ from the type strain of *B. cereus* by only one nucleotide. Yet phenotypic tests such as those in an API kit can separate the three organisms while reflecting the relationships implied. Similarly, comparisons of sequences from the two Antarctic psychrophiles *B. globisporus* and *B. psychrophilus* found over 99.5% 16S rRNA sequence identity yet the species had originally been distinguished by DNA relatedness as well as phenotypic characters.

Clostridium and relatives

Clostridium is an accumulation of well over 100 species, some of which are based upon relatively few and inadequately characterized strains, the criteria for admission being that they are sporulating anaerobes which do not reduce sulphate (this last excludes *Desulfotomaculum*). Members of the genus are diverse in their morphologies, which include coccoid, straight, curved, filamentous, and coiled rods (Fig. 11.4), and in their

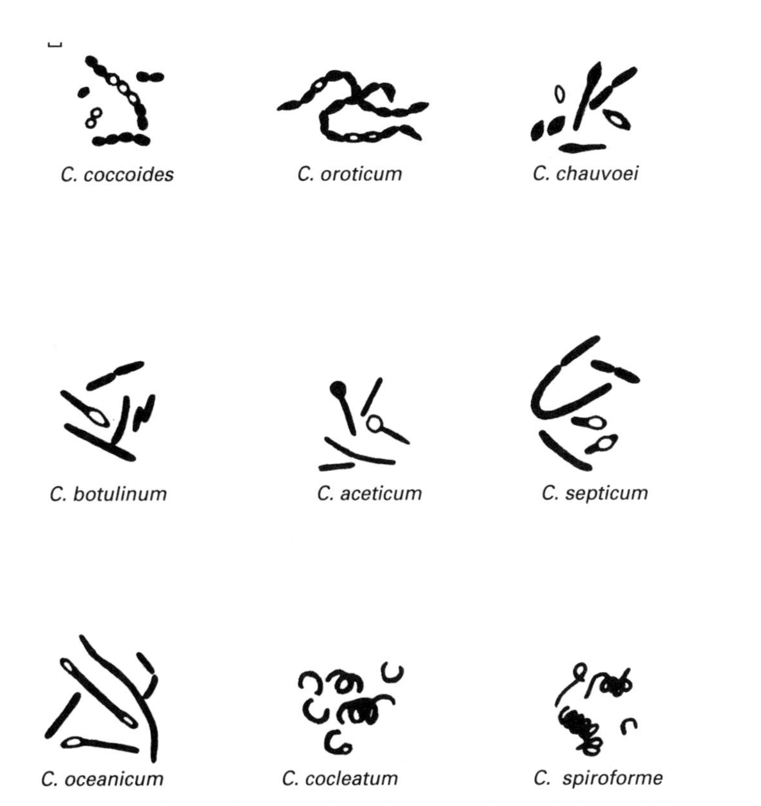

Fig. 11.4 Some examples to illustrate the wide range of morphological diversity seen amongst members of the genus *Clostridium*. Bar marker represents 1.0 µm.

nutritional requirements and metabolic activities, encompassing autotrophy, lithotrophy, organotrophy, and nitrogen fixation, and this is reflected in a wide range of mol% G+C. There have been no really comprehensive taxonomic studies of the genus and little recent work other than descriptions of numerous new species. Most numerical taxonomies of phenotypic characters, and chemotaxonomic and genomic analyses, have sought to clarify the structures of isolated groups of apparently related species. There has been more emphasis on the morphological and biochemical properties, toxicities and serological characters of the intensively studied pathogens as aids to identification.

Nucleic acid homology studies have, however, yielded some valuable insights. The largest investigation, covering 56 species, found little relatedness between species in the 41–45 mol% G+C range or with *Bacillus* species in the same range, but revealed three 23S rRNA homology groups among species in the 21–33 mol% G+C range; two of the groups were well defined and one was divisible into 10 subgroups. These groups and subgroups showed some correlation with certain phenotypic properties, including cell morphology, nutritional requirements, ability to ferment carbohydrates, products of fermentation and protein electrophoresis patterns, but not with others such as spore position, gelatin liquefaction, and ranges of carbohydrates and amino acids fermented. Some of these characters are useful in diagnostic keys but numerical analysis of published results shows poor congruence with the homology groups.

Toxins. Production of a toxin is not a reliable taxonomic character as it may be phage- or plasmid-mediated and the property can be transferred to other species or lost. *Clostridium botulinum* is a good example of a special purpose, artificial and monothetic classification (see Chapter 1). The species was identified solely on the basis of its strains producing the neurotoxin botulin, of which seven serologically distinct types, A–G, occur. However, these types do not necessarily correlate with other phenotypic characters, including PAGE analysis of cellular proteins, or with nucleic acid homologies, which readily allow the recognition of four subgroups containing toxigenic and non-toxigenic strains. These subgroups show higher relatedness to other *Clostridium* species than among themselves (Table 11.4), but *C. botulinum* long retained its single species status because of its members' clinical importance. The toxin type G strains and some non-toxigenic strains labelled *C. subterminale* and *C. hastiforme* have now been placed in the new species *C. argentinense*. *Clostridium perfringens*, however, has been divided into five types on the basis of toxins produced, but the species shows 100% nucleic acid homology and the toxin types cannot be differentiated on the basis of other phenotypic tests.

Phylogeny. Relatively few sequencing studies have included many representatives of *Clostridium* species, and much of our knowledge has been derived from surveys including clostridia as references. This information, along with the 23S rRNA hybridization study already mentioned, reveals six

Table 11.4 Subgroups of *Clostridium botulinum*

Metabolic type	Toxin types	Relatedness groups* DNA	23S rRNA	Related species
1 Proteolytic	ABF	I	I F	*C. sporogenes*
2 Saccharolytic	EBF	II	I A	*C. butyricum*
3 Saccharolytic	C†D	III	I H	*C. novyi*
4 Proteolytic	G‡	–	I K	

* DNA data from Lee & Riemann (1970) and 23S rRNA data from Johnson & Francis (1975).
† Indistinguishable from *C. sporogenes* by 16S rRNA restriction endonuclease patterns (Gurtler *et al.*, 1991).
‡ Reclassified as *C. argentinense.*

groups of *Clostridium* species. Groups I and II, which include the medically important species, are divided into 10 and two subgroups respectively. As with *Bacillus*, the analyses reveal several groups that contain mixtures of sporeformers and non-sporeformers, but the anaerobes show deeper branching overall and appear to represent more ancient lineages. In cataloguing studies several species show quite close relationships with species of *Eubacterium* (*C. barkeri, C. lituseburense*; groups IV and II-A), *Peptostreptococcus* (*C. lituseburense*) (both genera of convenience), *Sarcina* (*C. butyricum*; I-A), *Acetobacterium, Acetogenium* and *Thermoanaerobium* (the chemolithotroph *C. thermoaceticum*; V), and the anaerobic streptococci.

Comparisons of the few published 16S rRNA sequences show that *C. innocuum* and *C. ramosum* (both group VI) are related to the mollicutes (Fig. 11.1) (and *Erysipelothrix* according to catalogues); they appear to be no closer to the true clostridia than are *Bacillus, Lactobacillus, Staphylococcus* and *Streptococcus*. Again, as with *Bacillus*, a flexible approach is needed if a phylogenetically based classification is to be of practical value, and it has to be accepted that the clostridia will be distributed among several genera, families and orders as Cato & Stackebrandt (1989) have informally suggested (Table 11.3). Much sequencing work must be done before any such divisions can be made, but *C. bryantii* has already been transferred to a new genus, *Syntrophospora*, which is closely related to the asporogenous *Syntrophomonas wolfei* but not to any other Gram-positive organisms.

Comparisons of restriction endonuclease digest patterns of amplified 16S rRNA genes from medically important species revealed two groups of species, one containing *C. bifermentans, C. difficile* and *C. sordellii* (hybridization group II), the other *C. botulinum, C. perfringens, C. sporogenes* and *C. tetani* C and G (group I); although this approach analyses only a small part of the gene, and some strains appeared to have two 16S rRNA alleles (implied by variations of fragment occurrences within groups), the results were largely consistent with those from the 23S rRNA hybridization study. Notable differences were that *C. tetani* was previously assigned to group II and that *C. sporogenes* was indistinguishable from *C. botulinum* C.

Asporogenous rods

The lactic acid bacteria have long been of interest owing to their importance in the food, dairy and alcoholic beverage industries, and they were soon recognized as forming a natural group because of their many physiological and biochemical similarities. The largest and best-known genus in the group is *Lactobacillus*, and yet it is also one of the least satisfactorily classified. It was established in 1901 to accommodate non-sporulating, microaerophilic and fermentative Gram-positive rods and in 1919 Orla-Jensen, who pioneered the taxonomic application of physiological characters, divided the genus into the three groups or subgenera "*Betabacterium*" (heterofermentative — subsequently split by Garvie into active and inactive groups), "*Thermobacterium*" and "*Streptobacterium*" (both homofermentative) according to fermentation products and growth temperatures; three other lactic acid genera he recognized were *Streptococcus*, "*Betacoccus*" (now *Leuconostoc*) and "*Tetracoccus*" (now *Pediococcus*). The limited 16S rRNA sequencing data available show the last two genera to be closely related to *Lactobacillus* and it has been suggested that they be included in an expanded *Lactobacillus*, but for practical purposes their different morphologies argue against this. Remarkably, an arrangement resembling Orla-Jensen's division of *Lactobacillus* was still being used informally 67 years later in *Bergey's Manual of Systematic Bacteriology*; why should this be so?

Lactobacillus

Lactobacillus is shown by a G+C range of 32–53 mol%, low DNA relatedness between species, and some enzyme sequencing data to be an old and very heterogeneous genus. The rRNA sequence data suggest that most species branched off quite early, and over a short period as microaerophilic life on earth became possible. Given its importance the genus has been somewhat neglected by numerical taxonomists, with most studies concentrating on strains from special environments such as wine, cheese, French dry sausages, vacuum packed meats and Scotch whisky distilleries, rather than taking a broader view. The last and larger three of these studies were inconclusive however: the reference strains of different species not only failed to fit with subgenera or cluster with fresh isolates, but were also frequently separated only at quite high similarity levels (which would imply numerous new species), or else intermediate strains made recognition and practical distinction of taxa difficult. Moreover, in the last study, cluster memberships varied considerably as different similarity coefficients were used, and ordination implied still different groups, suggesting that clusterings had been random. On the basis of these large, but narrow, studies it appears that the genetic heterogeneity of the genus is not strongly reflected by the kinds of characters traditionally used in numerical taxonomies — perhaps owing to convergent evolutionary pressures or restricted genomic expression in particular environments.

Chemotaxonomic methods such as electrophoresis, immunology or assay of enzymes, cell wall amino acids, intracellular amino acid pools and whole-organism lipids have usually only been of value in differentiating between small groups of species or within certain of the subgenera, and some small studies appeared to show promise but were not followed up, and so overall these approaches have not become routinely applicable either. Consequently the arrangement in *Bergey's Manual of Systematic Bacterology* (1986) recognized three groups similar to those of Orla-Jensen, but based them upon fermentation products and omitted growth temperature and morphology (as many of the more recently described species would not fit the definitions of Orla-Jensen's groups): group I, obligately homofermentative — hexoses fermented to lactic acid by Embden–Meyerhof pathway, pentoses and gluconate not fermented (contains the thermobacteria); group II, facultatively heterofermentative — hexoses fermented to lactic acid by Embden–Meyerhof pathway, or by some species at least to lactic, acetic and formic acids and ethanol under glucose limitation, and pentoses fermented to lactic and acetic acids (contains the streptobacteria); group III, obligately heterofermentative — hexoses fermented to lactic acid, acetic acid (ethanol) and CO_2, pentoses to lactic and acetic acids, and hexose heterofermenters possessing other pathways (i.e. *L. bifermentans*, *L. divergens*) also included (contains the betabacteria).

These groups are of some value in identification, but they are not clearly recognizable from the rRNA sequence data published to date (Fig. 11.5) as most of the species show great phylogenetic depth. This depth means that many species within the groups show low DNA relatedness to one another.

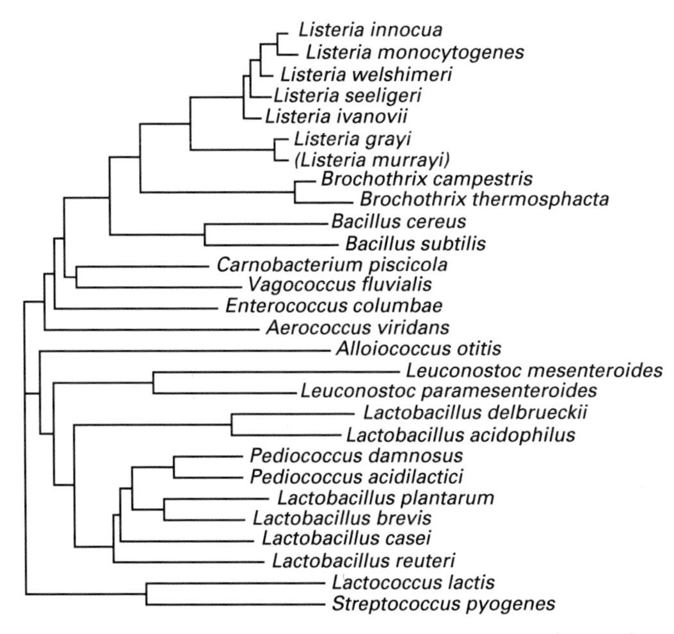

Fig. 11.5 An unrooted tree based upon 16S rRNA sequencing studies, indicating the evolutionary relationships between *Lactobacillus* species, *Listeria* species, and other Gram-positive bacteria.

Where they are high some taxonomic revisions of species and subspecies complexes have been possible: thus, *L. bulgaricus*, *L. delbrueckii*, *L. lactis* and *L. leichmannii* show more than 80% relatedness and only minor phenotypic differences so that they have been reduced to subspecies of *L. delbrueckii*, with *L. leichmannii* being regarded as a synonym of *L. lactis*. On the other hand *L. acidophilus* has been divided into six genomic species, most of which can be differentiated using phenotypic characters. In group II three complexes centred on *L. casei*, *L. plantarum* and *L. sake* with *L. curvatus* were recognized, yet none of the species outside these showed any specific relationships.

Although most species of group III fall within a range of only 6 mol% G+C, they show little DNA relatedness to each other and the phylogeny of the group is unclear. Several species such as *L. confusus* and *L. viridescens* have Lys–Ala–Ser or similar peptidoglycan types characteristic of *Leuconostoc*, rather than the Lys–Asp form typical of lactobacilli. Although the atypical species *L. bifermentans* does homoferment glucose to lactic acid, it may then split it into acetic acid, CO_2 and H_2, and rRNA catalogues and DNA–rRNA hybridization show a close relationship with *L. casei* (group II). Another species producing gas from glucose, *L. divergens*, has a *meso*-DAP direct peptidoglycan and a polyphasic study of it and several other atypical lactobacilli resulted in the proposal of the new genus *Carnobacterium*. *Carnobacterium* is quite closely related to another new genus, *Vagococcus*, which contains some motile, Lancefield group N streptococci that do not belong to *Lactococcus* (Fig. 11.5).

Listeria

For many years *Listeria* was monospecific and thought to be associated with the corynebacteria, and in the eighth edition of *Bergey's Manual* (1974) this genus, and three of its four species, were listed as of uncertain affiliation. It had been proposed in the same year, on the basis of DNA relatedness, that the closely related *L. grayi* and *L. murrayi* should be subspecies of a new taxon *Murraya grayi*, but this has not yet gained acceptance. However, biochemical properties, serology, base composition and DNA homologies clearly indicated that *L. denitrificans* belonged elsewhere and a new genus, *Jonesia* (now classified with the actinomycetes), was eventually created for it. Since 1974 the genus has attracted much attention from taxonomists (and subsequently from many others as the importance of *L. monocytogenes* became more widely appreciated), and four further species have been established.

Numerical taxonomy also found *L. grayi* and *L. murrayi* to be closely related and suggested reduction to synonymy, but supported its retention in *Listeria*, as did the data from cell wall, cytochrome, menaquinone and fatty acid analyses. Multilocus enzyme electrophoresis, DNA relatedness, rRNA gene restriction patterns and 16S rRNA sequence comparisons also strongly support the synonymy of the species, the last showing them as forming a distinct line of descent (Fig. 11.5), and their reduction to the single species *L. grayi* has been formally proposed. Other DNA relatedness studies not only supported the recognition of a separate genus for *L. grayi* but also revealed five groups of *L. monocytogenes* strains: one contained the type strain; one comprised strongly β-haemolytic strains of serovar 5 and was named *L. ivanovii*; non-pathogenic and non-haemolytic strains which belonged to serovar 6 became *L. innocua*; and the remaining two were designated *L. seeligeri* and *L. welshimeri*. The distinctness of these species is also evident from 16S rRNA comparisons. Although few phenotypic characters are known for distinguishing these species, they are readily divisible by multilocus enzyme electrophoresis. A numerical taxonomy found seven clusters of which five represented or probably represented the known species, while one of the others appeared to represent a new species.

From an earlier numerical taxonomy it was concluded that the *Lactobacillaceae* should contain *Erysipelothrix, Gemella, Lactobacillus, Listeria, Streptococcus* and another distinct taxon, worthy of genus rank, comprising strains of *Microbacterium thermosphactum*. For this last taxon, a catalase-positive rod, the new genus *Brochothrix* was later proposed. The closeness of *Listeria* and *Brochothrix* is confirmed by rRNA sequence comparisons (Fig. 11.5), which also support the inclusion of the two genera in the *Listeriaceae*, a family lying closer to *Bacillus, Carnobacterium* and *Enterococcus* than to *Lactobacillus*.

Important species

Endospores are structures formed within the cell in response to nutrient deprivation and they are usually oval, spherical or cylindrical, but sometimes kidney- or pear-shaped and may swell the sporangium (Figs 11.2 & 11.4). They are highly resistant to lethal agents such as radiation, heat, desiccation, and disinfectants, and are most important to the organisms' ecologies by allowing them to lie dormant for very long periods, perhaps 1000 years or more. The protoplast forms the core of an endospore and contains calcium dipicolinate, it is surrounded in turn by a membrane, a thick peptidoglycan cortex and a rigid proteinaceous outer coat. Endospores, and their formation, structure, function, germination and outgrowth, have been studied intensively for many years as models of prokaryote differentiation and because of their properties, but their resistance has only been partly explained in terms of the impermeability of the spore coat, dehydration of the core and protective proteins that bind to DNA.

Many endosporeformers appear to be ubiquitous, but the presence of spores in a particular environment does not necessarily indicate that growth occurs there, and this complicates ecological studies. Soil and other environments rich in organic matter are the primary habitats of these organisms, where they contribute to the degradation of organic material and the biological cycling of carbon and nitrogen, from where they may contaminate anything. They are consequently important in food spoilage, biodeterioration, intoxications and opportunistic infections of humans and other animals, and several species are true pathogens; some are obligately so.

As well as being members of the soil flora, *Bacillus* species are important in the medical, veterinary and industrial fields. Anthrax has been known since antiquity and remains a problem in many parts of the world. *Bacillus cereus*, a regular cause of food-borne diarrhoeal illness and an emetic type of intoxication food poisoning, is also an opportunistic pathogen, with bovine mastitis and human ophthalmitis being particularly serious. *Bacillus licheniformis* has been associated with bovine and ovine abortion and it, and other species, have caused various infections and food-borne illness. Of the species pathogenic for insects, *B. larvae* harms honey bees but *B. thuringiensis* and *B. popilliae* are widely used as insecticides; some *B. sphaericus* strains also have potential as insecticides.

Various species are involved in fermentations for the production of foodstuffs, such as cocoa, whereas *B. cereus* and *B. subtilis*, for example, are important in the spoilage of dairy and bakery products respectively. Members of the genus are also rich sources of extracellular enzymes, especially proteases (used in detergents) and amylases, and of peptide antibiotics such as bacitracin and polymyxin.

Clostridia are abundant in soil, intestinal tracts and other environments that are rich in decaying plant and animal materials and readily become anaerobic, such as sediments and sewage. They can also spoil meats, vegetables and cheese. Their potential in solvent production is largely unrealized because of the availability of petrochemicals and their main industrial importance is in the digestion of sewage sludge. Most attention, however, has been focused on those species that are opportunistic or true pathogens, and on the toxins involved. Around 60 species have been isolated from human and animal specimens but less than one-third of these are regularly encountered and the clinical significances of many are uncertain. With some important exceptions such as tetanus (caused by *C. tetani*), food-borne illness caused by *C. perfringens*, and intoxication by ingestion of preformed *C. botulinum* toxin, infections are generally acquired endogenously. *Clostridium perfringens* is considered to be a true pathogen and causes a wide variety of infections — it is the principal agent of gas gangrene, followed in importance by *C. novyi*, *C. septicum*, *C. histolyticum*, *C. bifermentans* and others — and also causes enterotoxaemias in animals. Many clostridial infections are connected with trauma and impaired blood supply, but some, such as infant botulism, antibiotic-related gastro-intestinal illness caused by *C. difficile*, enterotoxaemias, and braxy in sheep (caused by *C. septicum*), are associated with gut colonization.

Lactobacilli are usually acid-tolerant or acidophilic and are found in many environments, where their production of lactic acid commonly reduces the pH to a level that is inhibitory to most competitors. They are found in plants, fruit juices, milk and dairy products, alcoholic beverages, meat and meat products and fish, where they may be flavour-producing or spoilage organisms. While the definition of these is largely a matter of preference, lactobacilli are most important in food technology and contribute to the manufacture of silage, sauerkraut, pickles, wine, cider, sourdough, yoghurt, cheese, fermented sausages and many other products. Some species or subspecies have adapted to special niches and are rarely found outside these — examples are *L. delbrueckii* subsp. *bulgaricus* in yoghurt and cheese and *L. sanfrancisco* in sourdough. They are also (often valuable) members of the natural flora of the mouth, gut and vagina of many homothermic animals, including humans. In the upper gastrointestinal tract of certain young animals, and in the vagina, they are symbiotic and protect against harmful infections. Although lactobacilli are generally considered beneficial (there is interest in using some species as probiotics) and non-pathogenic, a few species have been implicated in endocarditis and several septic infections, and *Carnobacterium* species can be fish pathogens.

Listeria monocytogenes, long known as a pathogen of humans and other animals, gained notoriety in the late 1980s as its ability to grow in refrigerated foods and survive reheating became more widely appreciated and as outbreaks of infection received wide publicity. Humans most at risk of infection, which is usually meningitic and sometimes includes septicaemia, are neonates, the old, and those compromised by underlying illness, immunosuppressive therapy or pregnancy.

Identification

Most *Bacillus* species are easily cultivated and several are regularly encountered, producing very variable colonies that are distinctive none the less. Many species may be selectively enriched, and differential selective media are available for some; *B. cereus*, for example.

Sporulation is a valuable character for identification to genus and species level but it is too often overlooked. Species are differentiated by patterns of results in traditional cultural, microscopic, physiological and biochemical characters or API test kits (Fig. 2.1) and chemotaxonomic approaches based upon pyrolysis have also been used. Because of the perceived importance of *B. anthracis* as a potential biological weapon, there has been intensive study of PMS, immunofluorescence and flow cytometry for identification. Flagellar serotyping schemes for *B. cereus* and *B. thuringiensis* are well established.

Isolation of clostridia relies on anaerobic methods and various enrichment procedures may be used; also, spores may be selected for by pasteurization or ethanol treatment. Some species do not sporulate readily, and cell shapes and Gram reactions are variable, so that tests may be necessary to exclude other genera of anaerobes. None the less, the clostridia most often encountered in clinical specimens are readily identified from cultural and microscopic characters, and tetanus is recognized from the symptoms it causes. Fluorescent-antibody staining is useful for identifying *C. chauvoei* (which causes herbivore infections), *C. novyi*, *C. septicum* and *C. sordellii*. *Clostridium botulinum* and *C. difficile* may be detected by toxin testing.

Identification of other species relies upon cultural and microscopic observations (particularly sporangial morphology) and relatively few biochemical tests (especially for proteolytic and saccharolytic activities). Several miniaturized test kits are commercially available. Gas–liquid chromatography of the products of carbohydrate and amino acid metabolism is useful (or of cell wall fatty acids, as used in a commercial system), especially for the rarer species, but is not always as valuable as it is for other anaerobes because profiles can be inconsistent. None the less, for *C. difficile* GC of culture headspace offers advantages of accuracy and speed.

The lactic acid bacteria are metabolically very similar to each other and some genera and species show a morphological range from rods to cocci, making identification even to genus level difficult; PCR amplification of rDNA followed by nucleic acid probing promises to be of help. Simple conventional methods such as catalase activity and gas from glucose are still valuable for differentiation, and routine biochemical test kits are available for *Lactobacillus* and *Listeria*, but chemotaxonomic methods such as enzymic determination of lactic acid isomers produced, cell wall analysis by TLC, GLC for teichoic acids, and PAGE for lactic acid dehydrogenase mobilities are also useful for separating *Lactobacillus* species.

12 The Gram-positive cocci

The story of the streptococci is the history of their classification. In the early days of the discovery of these organisms in wounds and septic processes ... the wideness of their distribution was unknown and the possibility of more than one type of organism existing under a single morphological garb was scarcely considered.

J.H. Dible, 1929, *Recent Advances in Bacteriology*

Introduction

The 19 genera grouped together as the Gram-positive cocci are chemoorganotrophic, mesophilic and non-sporulating, but have few other characters in common overall. As well as being phenotypically diverse they represent several evolutionary lines; the family *Micrococcaceae* brought together *Micrococcus, Stomatococcus, Staphylococcus* and *Planococcus* which belong to four quite different phylogenetic groups (*Micrococcaceae* is now an actinomycete family — see Chapter 14). The main characters of each of the genera are shown in Table 12.1, and Fig. 12.1 gives a schematic representation of the evolutionary pathways, as inferred from very limited data. As mentioned in Chapter 11, the placing of the Gram-positive rods and cocci in separate chapters is purely for convenience.

The 19 genera may be divided into aerobes, facultative anaerobes and microaerophiles, and anaerobes, or into two groups according to whether or not they produce catalase and/or cytochromes; *Alloiococcus, Deinococcus, Micrococcus, Planococcus, Staphylococcus* and *Stomatococcus* are all positive whereas all other genera are catalase-negative or weakly positive and have only a few species that may produce cytochromes when grown aerobically with haemin. However, oxygen relations and environmental niches do not correlate with evolutionary pathways.

Deinococcus, Micrococcus and *Stomatococcus*

Deinococcaceae is a monogeneric family of radiation and desiccation resistant organisms that produce red carotenoid pigments and which were classified in *Micrococcus* for many years, until non-routine characterization tests showed that they represent an unrelated group. The family rank is to emphasize the distinctness of the organisms and provide a home for relatives as and when they are discovered — much the same as with the *Legionellaceae*. Although deinococci are Gram-positive, 16S rRNA cataloguing revealed no phylogenetic alignment with any other Gram-positive organisms or other eubacteria but indicated a distant relationship with very different organisms — *Thermus* species, which are Gram-negative, rod-shaped bacteria that inhabit hot springs and other heated waters, and which, it is interesting to note, commonly produce carotenoid pigments.

Table 12.1 Characters of the Gram-positive cocci

Genus	Number of species	Main habitats	Mol% G+C	Gaseous requirements	Motility	Catalase	Cyto-chromes	Glycine in interpeptide bridges
Deinococcus	4	Unknown	62–70	Aerobic	–	+	+	+
Micrococcus	9	Mammalian skin	64–75	Aerobic	–	+	+	–
Planococcus	2	Marine	39–52	Aerobic	+	+	+	–
Alloiococcus	1	Human ear	44–45	Aerobic	–	+	ND	ND
Staphylococcus	31	Humans, animals and environment	30–39	Facultative	–	+	+	+
Stomatococcus	1	Human mouth	56–64	Facultative	–	w/–	+	+/–
Streptococcus	45	Humans and animals	34–46	Facultative	–	–	–	+
Enterococcus	16	Human and animal intestines, and environment	37–45	Facultative	–	–	–	+
Lactococcus	4	Milk products	38–42	Facultative	–	–	–	+
Vagococcus	2	Animals and water	33–37	Facultative	+	–	–	–
Leuconostoc	10	Plants and milk products	37–45	Facultative	–	–	–	–
Pediococcus	8	Beer, plants and milk products	34–42	Facultative	–	–	–	–
Aerococcus	1	Hospitals and lobsters	35–40	Microaerophilic	–	w/–	–	No bridge
Gemella	1	Humans	32–35	Aerobic/ facultative	–	–	(–)	–
Peptococcus	1	Humans	50	Anaerobic	–	–/w	–	–
Peptostrepto-coccus	11	Humans	27–45	Anaerobic	–	–/w	–	+
Ruminococcus	8	Rumen and large bowel	39–46	Anaerobic	–	–	ND	No bridge
Coprococcus	3	Gut contents	39–42	Anaerobic	–	–	ND	No bridge
Sarcina	2	Gut contents	28–31	Anaerobic	–	–	ND	+

ND, not determined; w, weak.

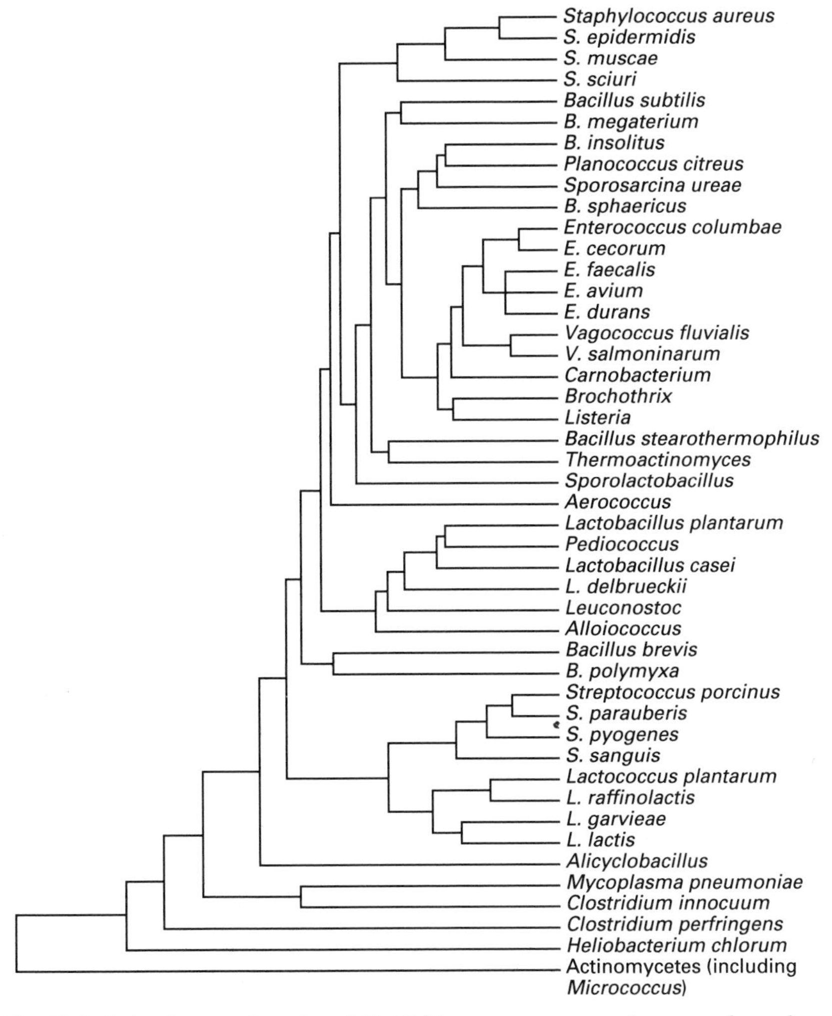

Fig. 12.1 A dendrogram based on 16S rRNA sequence comparisons, to show the probable evolutionary relationships between the main groups of Gram-positive cocci.

Sequencing of 16S rRNA has confirmed the relationship and supports the recognition of a *Deinococcus–Thermus* phylum.

Some other thermophiles such as *Thermotoga* and *Thermomicrobium* show appreciable sequence similarity to *Thermus,* but this is believed to be convergence because of the high G+C contents of thermophiles' rRNAs. Deinococci also differ from the other cocci in phenotypic properties: their cell walls are atypical for Gram-positive bacteria as they have an outer membrane of protein and lipid, and they have a high proportion of even-numbered, straight-chain saturated and unsaturated fatty acids (principally palmitoleate), both features more characteristic of Gram-negative bacteria. Their phospholipids are also unusual, as phosphatidylglycerol and its derivatives are absent.

It was suggested that newcomers to the group might not be Gram-positive or coccoid, the *Deinococcus* Gram reaction presumably being due to an unusually thick peptidoglycan, and indeed Gram-negative, rod-shaped, radiation-resistant organisms were subsequently described. They produce pink or red pigments, and have cell wall structures and other chemotaxonomic features similar to *Deinococcus* and have been called *Deinobacter grandis.*

The other five catalase-positive genera all belong to the Gram-positive phylum, but again correlations with oxygen relations, catalase and cytochromes are absent. The Gram-positive bacteria comprise two distinct subdivisions within their phylum; the high G+C one (which appears to be much younger than the other, low G+C one) contains actinomycetes and their relatives and includes the micrococci and stomatococci. Stomatococci appear to represent a quite distinct line of descent according to 16S rRNA sequencing analysis and DNA–rRNA hybridizations and several phenotypic characters such as their thick capsules, poor growth on nutrient agar, firmly adherent growth on richer media and unusual peptidoglycan (having a single monocarboxylic amino acid for the interpeptide bridge) differentiate them from other cocci.

Micrococcus is most closely related to the corynebacteria–arthrobacter group, and is not related to *Staphylococcus*. *Arthrobacter* species typically have a life cycle including a coccal stage, and as micrococci cannot be separated from them by phylogenetic data it is believed that the micrococci represent degenerate arthrobacters that are locked in this coccoid stage. Other characters support this idea: their peptidoglycans both contain lysine as the diamino acid but have no glycine, they have similar phospholipids, and their menaquinones, containing 7–9 isoprene units in the case of *Micrococcus* and 8–9 in *Arthrobacter*, are often hydrogenated.

Members of the 'clostridial' subdivision

The remaining 16 genera all belong to the low G+C or 'clostridial' subdivision which shows quite deep branching and so is considered to be ancient in evolutionary terms. Limited rRNA sequencing work indicates five or more main clusters (see Table 11.3); four contain clostridia and relatives (including the rapidly evolving mollicutes), and one contains *Bacillus*, *Lactobacillus*, *Staphylococcus* and *Streptococcus* which are, interestingly, all aerobic, facultative or aerotolerant organisms. It has been suggested that the emergence of this cluster may correspond with the change of the earth's atmosphere from anaerobic through microaerobic to aerobic. *Pediococcus*, *Staphylococcus* and *Streptococcus* are all related to the *Bacillus* cluster, and *Planococcus* shows a particularly close relationship with the coccus *Sporosarcina* and the closely related rod *B. pasteurii.* They have similar cell wall chemistries and phospholipid patterns and are motile; it should be noted that the concept of motile cocci was derided for many years for no apparent reason.

Of the facultative and microaerophilic, catalase-negative cocci, *Aerococcus*, *Gemella* and *Pediococcus* showed a close relationship with *Streptococcus* in a numerical taxonomic study, and oligonucleotide cataloguing has shown this to hold from the evolutionary point of view, with *Aerococcus* and

Gemella representing independent lines of descent within the group. *Leuconostoc*, however, lies closer to the obligately heterofermentative species of *Lactobacillus*, with which it represents an evolutionary branch distinct from the other lactobacilli (Fig. 12.1). Too few of the anaerobic genera have been adequately described for a clear phylogenetic picture to emerge, but they all appear to be offshoots of several clostridial groups (see Table 11.3). What is clear is that *Peptostreptococcus* is a very heterogeneous and probably artificial taxon and that it should be regarded as a temporary genus.

Staphylococcus

The pre-eminent species of this genus, *S. aureus*, so called because of its golden pigment, has been recognized as an important human pathogen for well over a century; it and the *apparently* harmless species *S. epidermidis* were the only species of the genus until 1974 when *Micrococcus saprophyticus*, which has a primarily respiratory metabolism but little else in common with the micrococci, was transferred. Subsequently, nucleic acid hybridization and chemotaxonomic studies have resulted in the description of a further 28 species and seven subspecies.

All species are parasites of warm-blooded animals and some show adaptation to particular hosts. *Staphylococcus aureus* is exceptionally widespread and although primates seem to be the major natural hosts it can be isolated from most animals including marine ones, whereas, for example, *S. caprae*, *S. equorum*, *S. gallinarum* and *S. intermedius* have particular associations with goats, horses, chickens and dogs respectively. *Staphylococcus delphini* was isolated from dolphins, and *S. felis* from cats and dogs.

It is possible to differentiate between strains of the same species from different hosts by DNA relatedness; thus, there are human and non-human primate ecovars of *S. warneri* and *S. auricularis*. Human species, furthermore, may show preferences for different niches on the body, and these seem to correlate with variations in resistance to host secretions, moisture and light. Examples are *S. capitis* and *S. auricularis*, which are mainly found on the head and ears, and *S. aureus* which is mostly carried in the nose, and while *S. epidermidis* and *S. hominis* both colonize the axillae and pubic areas heavily, the latter species is the more successful on drier areas such as the limbs. As already indicated, it is the genomic and chemotaxonomic methods, that is to say approaches used for phylogenetic studies, that have enabled the recognition of so many species and subspecies, and this supports the belief that **conjugate evolution** has occurred (the staphylococci evolving together with their hosts), with the organisms adapting to changing skin habitats. This is particularly so in humans given the large number of species they carry.

In the mid-1970s analyses of cell walls and many biochemical tests and other phenotypic characters were used to define the three existing species better and describe seven new species, and it was possible to recognize two species complexes — the novobiocin-resistant species and the other coagulase-negative species. It was also found that some species showed special associations with particular habitats, but these were never sufficiently distinctive or consistent to be reliable for identification purposes. Subsequently, DNA homologies and thermal stabilities of hybrid DNAs confirmed these species, played a large part in the recognition of some 20 further species, and allowed the recognition of six species groups in agreement with those complexes recognized on the basis of phenotypic tests (Table 12.2). It can be seen that not all species are accommodated within these groups, and this may be taken to indicate earlier branchings or more rapid evolution within habitats.

The groupings are also supported by a wide diversity of phenotypic characters including cell wall peptidoglycan structure and teichoic acid type, menaquinone composition, cellular fatty acids, class of fructose-1,6-bisphosphate aldolase, and novobiocin resistance, whereas other characters such as lactate isomers produced from glucose fermentation, and routine tests for hydrolytic enzymes, acid from sugars, colonial size and pigmentation etc., are more useful for separating species within groups, and for routine identification, without contradicting the groupings (Table 12.2). This is not to say that these species groups actually represent single specie — stringent DNA hybridizations and numerous other characters (such as those useful for identification) indicate otherwise.

At a higher level, DNA–rRNA hybridization and 16S rRNA sequencing show that the staphylococci form a very coherent group and remove doubts, arising from numerical taxonomic studies, about the membership of the species *S. lentus* and *S. sciuri,* which appear to represent early branching (Fig. 12.2). Members of the genus, as a whole, show a DNA base range, cell wall and fatty acid compositions, and cytochrome patterns which set them apart from other genera. Some of these and other characters are useful in separating them from other Gram-positive cocci in the routine laboratory. The nearest relatives of the staphylococci appear to be *Bacillus, Brochothrix thermosphacta, Enterococcus* and *Planococcus* and it has been suggested that the genus be placed in a new family, *Staphylococcaceae.*

Streptococcus and relatives

Rosenbach proposed the species *Staphylococcus aureus* and *Streptococcus pyogenes*, the type species of their respective genera, in the same publication in 1884, but while the staphylococci remained taxonomically static for many years, the great diversity of the streptococci was recognized early and attempts to classify these organisms satisfactorily were soon made. It is therefore rather surprising that the taxonomy of the group (for homogeneous genus it is not) is still unsatisfactory. This is explained by the apparent success of early classification/identification schemes and the importance of different members of the genus to such disparate groups as medical bacteriologists, veterinarians and dairy microbiologists who usually limited their classification studies to the species of particular interest to themselves.

Table 12.2 Some characters of the staphylococci

Species	DNA homology group*	Cell wall peptido-glycan type†	Mena-quinone type‡					FBP aldolase§ class	Lactate type produced¶		Novobiocin resistance	Coagulase	Urease
			5	6	7	8	9		D	L			
S. aureus		LG_5			w	+	w	I	+	+	–	+	w**
S. auricularis		LG_5 (LG_4S)						I	w		–	–	–
S. epidermidis	1	LG_4S		w	+			I		+	–	–	+
S. capitis	1	LG_4S		w	+			I		+	–	–	–
S. warneri	1	LG_4S		w	+			I	+	+	–	–	+
S. saccharolyticus	1	LG_4S						I	w	w	–	–	
S. caprae	1	LG_4S		w	+	w		I		+	–	–	+
S. hominis	1	LG_4S		w	+			I	+		–	–	+
S. haemolyticus	1	LG_4S		w	+			I	+		–	–	–
S. cohnii	2	LG_5		w	+			I	w	w	+	–	–
S. xylosus	2	LG_5		w	+			I	w	w	+	–	+
S. saprophyticus	2	LG_4S		w	+			I	w	w	+	–	+
S. gallinarum	2	LG_5						I		+	+	–	+
S. kloosii	2	LG_5 (LG_4S)			+					w	+	–	d**
S. equorum	2	LG_5		+	w					w	+	–	+
S. arlettae	2	LG_5			+					w	+	–	–
S. carnosus	3	LG_5			+			I	+	+	–	–	–
S. simulans	3	LG_5		w	+			I	w	+	–	–	+
S. intermedius	4	LG_4S		w	+			I and II		+	–	+	+
S. delphini	4	LG_5			+						–	+	
S. hyicus	5	$LG_4(S)$		+	w			I and II		+	–	(+)	d
S. chromogenes	5	$LG_4(S)$			+			I and II		+	–	–	d
S. sciuri	6	LAG_4	w	+				I		w	+	–	–
S. lentus	6	LAG_4	w	+				I	w	w	+	–	–
S. caseolyticus		LG_5						II		w	–	–	–
S. lugdunensis		LG_5			w	+	w				–	–	d
S. schleiferi		$LG_4(S)$			+	w					–	–	–
S. felis											–	–	+
S. muscae		LG_5 LG_4S									–	–	–
S. piscifermentans		LG_4		w	+	w			+	+	–	–	+
S. pasteuri		LG_4S							+	+	–	–	+

* The numbers are allocated arbitrarily.
† LAG_4, Lys-Ala-Gly_{4-5}; LG_4, Lys-Gly_{4-5}; LG_4S, Lys-Gly_{4-5}, Ser; LG_5, Lys-Gly_{5-6}; minor components shown in parentheses.
‡ Numbers refer to chain lengths; w, minor component.
§ FBP, fructose-1,6-bisphosphate.
¶ Configuration of lactate produced from glucose fermentation; w = trace.
** w, weak; d, results differ between strains.

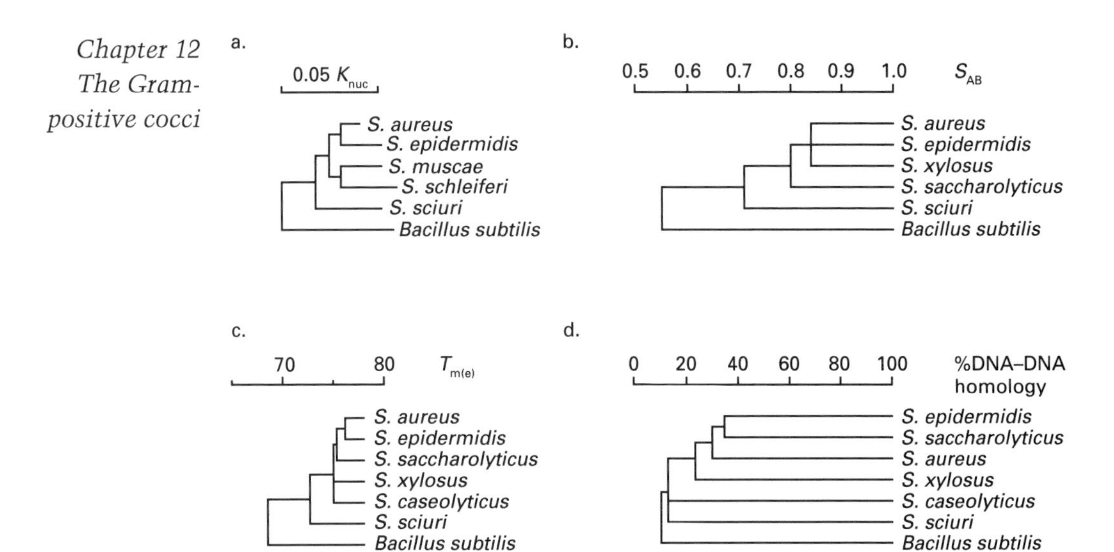

Fig. 12.2 Phylogenetic relationships among some members of the genus *Staphylococcus* based upon: a, 16S rRNA sequencing; b, 16S rRNA oligonucleotide cataloguing; c, DNA–rRNA hybridization; d, DNA–DNA homology. (a) adapted from Hájek *et al.* (1992).

The earliest studies, however, were quite wide-ranging: Andrewes & Horder used a variety of morphological observations and biochemical tests to characterize over 1000 isolates from humans, milk and air, and recognized eight groups in 1906; Orla-Jensen used a wider range of tests in a study of mainly dairy organisms and recognized 10 groups in 1919, the pathogenic strains being lumped as *S. pyogenes* for want of fuller information. There was much activity between the 1900s and 1930s as tests were further developed or new ones devised, but little taxonomic progress was made. Then in 1937 Sherman suggested that initial segregation of strains into groups according to growth temperatures, reducing ability and resistance to heat, alkalinity and methylene blue, would make interpretation of fermentation and other tests more intelligible, and he divided his 20 species (isolates of which had come from a variety of habitats) into the four groups pyogenic, viridans, lactic and enterococcus but excluded the pneumococcus and all the strict anaerobes from the genus (Table 12.3). His exclusion of the anaerobes was vindicated some 50 years later by genomic studies, and for the other species it is almost incredible that his groupings and the tests on which they were based have largely endured to the present day.

The group-specific nature of polysaccharide haptens in the cell walls of many of the β-haemolytic, pathogenic streptococci was recognized by Lancefield in the early 1930s and it was soon shown that her precipitin test groups correlated with the species recognized by Sherman's methods; so began an approach to classification that, because it relied on one small group of chemotaxonomic markers, led ultimately to confusion. Initially, species were defined on the basis of physiological and serological characters, and it was appreciated that the serological groups C, D and N each contained more than one species or biotype, but many of the other serological groups, some 15 of them, were not adequately characterized by other methods and yet they were often taken to represent distinct species. It gradually became apparent that the correlation between serogroup and physiological characters was not holding, that strains of some well-defined species may carry different Lancefield antigens, that the same antigen may be found in more than one species, and that very few species are represented by a distinct serogroup. Even *S. pyogenes*, a name often taken as synonymous with Lancefield type A, is not the only streptococcus to carry this antigen.

It must also be remembered that no group specific antigen was known for many of the streptococci, and this prompted further efforts to detect them; where successful, however, the results were often unhelpful as group antigens were found in species that were not shown to be closely related by other methods. For example, the group D antigen which was associated with the enterococci was also found in *S. bovis* and *S. equinus* (two viridans species), and this led to argument about their correct taxonomic positions. It is of interest that the group D and N (associated with the 'lactic' species) antigens are, in fact, glycerol teichoic acids located between cell wall and membrane, whereas the other Lancefield antigens are rhamnose-containing polysaccharides of the cell wall.

The emphasis on serological classification resulted not so much in an absence of other taxonomic studies, but in a neglect of their implications. Approaches included numerical taxonomy, resistances to antibiotics and other chemicals, cell structure, peptidoglycan structure, cell wall carbohydrates, teichoic acids and amino acids, protein homologies, gel electrophoresis, biochemical pathways, electron transport systems, plasmid transfer and DNA relatedness. Such work burgeoned in the 1970s as the shortcomings of Lancefield typing for taxonomy became apparent, but as Jones pointed out in 1978, the majority of these studies were very limited in the range of species included and so did not give a broad idea of relationships within and without the genus. Thus Jones divided the genus into seven artificial groups (pyogenic, pneumococci, oral, faecal, lactic, anaerobic and other) in order to allow review of these many and mostly restricted taxonomic studies.

In 1983 Bridge & Sneath published a numerical taxonomy of over 200 strains which attempted to cover the entire genus. They recognized 21 species and eight other phenons in three main clusters: the enterococci, which were well separated from the other groups, and then two less distinctly separated clusters containing in one the viridans, lactic, thermophilic and pneumococcal species groups, and in the other the pyogenic species group. As the viridans species fell into two clusters a paraviridans group was recognized, and the pyogenic group was similarly divided (Table 12.3). It was also found that *Aerococcus*, *Gemella*, *Leuconostoc* and *Pediococcus* were close relatives of the streptococci. This work should have enabled the interpretation of data from the many other, smaller studies but further developments at this time and subsequently were chiefly in the area of genotypic characterization.

Table 12.3 Developments in the classification of the streptococci

1937, Sherman (Physiological tests)		1983, Bridge & Sneath (Numerical taxonomy)		1987, Schleifer, Kilpper–Bälz and others (Genotypic and chemotaxonomic)	
*PYOGENIC	†(9)	PYOGENIC	(5)	PYOGENIC	(5)
‡*S. pyogenes*	§A	*S. pyogenes*	A	*S. pyogenes*	A
S. equi	C	*S. equi*	C	*S. equi* subspp.	C
S. mastitidis	B	*S. agalactiae*	B	*S. dysgalactiae*	C
Lancefield	C, E, G, H	"*S. equisimilis*"	C	*S. canis*	C
		Human		*S. iniae*	
		Lancefield	B		
		PARAPYOGENIC	(3)		
		S. uberis	(E)		
		"*S. dysgalactiae*"	C		
		Lancefield	R, S, T		
				AGALACTIAE	(1)
				S. agalactiae	B
VIRIDANS	(4)	PARAVIRIDANS	(7)	ORAL	(14)
S. salivarius	(K)	*S. salivarius*	(K)	*S. salivarius*	(K)
S. equinus	D	*S. equinus*	D		
S. bovis and varieties	D	*S. bovis*	D		
S. thermophilus		*S. mutans*		*S. mutans* group	(6)
		VIRIDANS	(4)		
		S. oralis		*S. oralis*	*S. oralis* group (8)
		S. sanguis	H	*S. sanguis* H	
		S. mitis	H	*S. mitis* H	
		"*S. milleri*"	F, G, C, A	*S. anginosus* F ("*S. milleri*")	
		PNEUMOCOCCAL	(1)	*S. pneumoniae*	
LACTIC	(2)	LACTIC	(1)	LACTOCOCCUS	(4)
S. lactis	N	*S. lactis*	N	*Lactococcus lactis* subsp. *lactis*	N
S. cremoris	N			*L. lactis* subsp. *cremoris*	N
ENTEROCOCCUS	(4)	ENTEROCOCCAL	(4)	ENTEROCOCCUS	(9)
S. faecalis	D	*S. faecalis*	D	*Enterococcus faecalis*	D
S. durans	D	*S. faecium*	D	*E. faecium*	D
S. liquefaciens	D	"*S. avium*"	Q	*E. avium*	Q
S. zymogenes	D	*S. gallinarum*	D	*E. gallinarum*	D
		THERMOPHILIC	(1)	OTHER STREPTOCOCCI	(10)
		S. thermophilus		*S. thermophilus*	
				S. uberis	E
				S. equinus	D
				S. bovis	D

* Group.
† Number of species, and/or subgroups.
‡ Main species.
§ Lancefield group.

Relationships at the species level have been resolved by DNA homologies and subdivisions within the genus have been revealed by DNA–rRNA hybridization and 16S rRNA sequencing supported by chemotaxonomic data. In the mid-1980s therefore, Schleifer and others formally proposed the new genus *Enterococcus* to include the faecal streptococci, and several other Lancefield group D species were soon added. The lactic species were transferred to the new genus *Lactococcus*, and the anaerobic streptococci were shown not to belong to the genus, with *S. parvulus* and *S. morbillorum* being transferred to *Peptostreptococcus* and *Gemella* respectively, and the others lying closer to certain clostridia. Subsequently, 16S rRNA sequence analyses of motile, group N, *Lactococcus*-like strains from water and chicken faeces led to the new genus *Vagococcus* being proposed (Fig. 12.1).

These developments leave the genus *Streptococcus sensu stricto* with three groups — the pyogenic, oral and other streptococci — with *S. agalactiae* ungrouped, which presents a rather heterogeneous assemblage that may warrant the proposal of further new genera. Although DNA relatedness studies reveal that these groups are more closely related to each other than they are to *Enterococcus*, *Lactococcus* and the anaerobic cocci, they can be separated by the same methods and the oral group can be subdivided. The viridans or oral group has been a taxonomic headache for many years. Attempts were made to classify them by physiological and biochemical tests because they are not separable by Lancefield typing, but such work was hampered by a hopeless disarray in their nomenclature — something which emphasizes the vital importance of satisfactory schemes and attention to the *Bacteriological Code* for all bacteria. Species names such as "*S. viridans*" and "*S. mitior*" were widely used but had not been validly published, so that no type strains were available for reference purposes; as a result different workers sometimes applied different names to the same strains. Furthermore, several names were applied to heterogeneous collections or groups which bore other names; "*S. milleri*" for example, included *S. anginosus*, *S. constellatus* and *S. intermedius*, and *S. oralis* contained "*S. mitior*", *S. sanguis* and "*S. viridans*".

Nucleic acid hybridization studies have allowed this exceptionally confusing mess to be tidied up. By DNA–rRNA hybridization it has been shown that most oral streptococci form a distinct group, and DNA homologies have enabled the members of the "*S. milleri*" and *S. mutans* groups to be recognized as individual species and for *S. oralis* to be given a satisfactory emended description (it was subsequently suggested that *S. oralis* is a later synonym for *S. mitis*). In some cases chemotaxonomic markers have supported the conclusions; for example, a cluster analysis of enzyme electrophoretic types correlated well with the homology groupings, and *S. oralis* can be distinguished from other streptococci by the unique composition of its cell wall, which contains ribitol and choline (also found in the close relative *S. pneumoniae*) and has a directly cross-linked peptidoglycan. However, although routine phenotypic tests have been found to correlate with this genomically and chemotaxonomically defined species, they are not always consistent enough to be reliable, and this is not an unusual problem with the oral streptococci. Also, while *S. oralis* can be confidently identified by its cell wall composition it is worth noting that this character is not necessarily of value with other species as, for example, strains of the biochemically and genetically defined *S. salivarius* show two quite different peptidoglycan types.

Yet further problems arise with the untidy collection of unassigned species and groups called the 'other streptococci'. *Streptococcus bovis* and *S. equinus*, for example, are names applied to a heterogeneous selection of strains; one study divided them into six DNA relatedness groups and proposed two of these as the new species *S. alactolyticus* and *S. saccharolyticus*, yet the type strains of *S. bovis* and *S. equinus* are very closely related and occupied the same relatedness group. A clearer picture has emerged from a larger set of rRNA sequence comparisons (Fig. 12.3): these support the inclusion of *S. agalactiae* and several of the 'other streptococci' group in the pyogenic group, reveal two subgroups in the oralis group, show *S. thermophilus* to be related to *S. salivarius*, and find *S. bovis* and *S. equinus* to represent a distinct group.

Although Lancefield grouping is no longer favoured for classificatory purposes it remains most valuable for identification of the pyogenic species. As the medical and veterinary importances of many of the oral and 'other' groups of streptococci become better appreciated a major challenge is to find phenotypic characters for reliable identification systems to be prepared, for without such systems the taxonomic advances will not become established.

Important species

Although the staphylococci are parasites of the body surfaces of warm-blooded animals and their ability to cause disease is largely incidental, several are important pathogens of humans (especially hospital patients) and other animals. *Staphylococcus aureus*, in particular, carries a large armoury of virulence factors to overcome host defences and causes acute

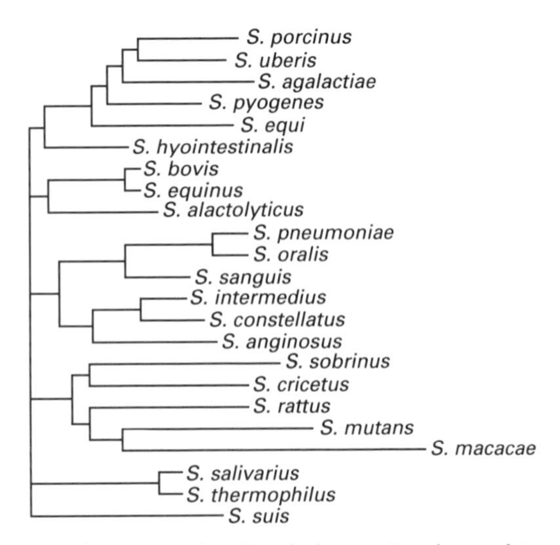

Fig. 12.3 Unrooted tree showing the phylogenetic relationships of some *Streptococcus* species. Adapted from Bentley *et al.* (1991).

infections ranging from localized and self-limiting to deep-seated and often life-threatening conditions; it also causes acute toxaemias such as staphylococcal food poisoning, toxic shock syndrome and scalded skin syndrome, and exudative epidermatitis of pigs, owing to specific protein toxins. It is now well recognized that the coagulase-negative staphylococci are also important causes of disease; *S. saprophyticus* is second only to *Escherichia coli* as a cause of urinary tract infections, and *S. epidermidis* is particularly associated with iatrogenic infections following all kinds of implant surgery. Multiple drug resistance is a particular problem in coagulase-positive and -negative species causing nosocomial (hospital-acquired) infections and in many clinical contexts such strains are now usual, and appear to evolve to take advantage of the niches provided by prosthetic surgery. Infections with epidemic hospital strains of *S. aureus* resistant to the β-lactamase-resistant penicillin methicillin (MRSA strains) and most other antibiotics can be very difficult to treat. The three species of greatest importance in veterinary medicine are *S. aureus*, *S. intermedius* and *S. hyicus*, which cause mastitis, dermatitis, abscesses, arthritis, osteomyelitis and septicaemia of a variety of domestic animals.

As befits their greater diversity, the streptococci are more wide-ranging in their pathogenic and other effects than the staphylococci. Clinically speaking, they are commonly artificially divided into four main groups:

1 β-Haemolytic, Lancefield-groupable species: the most pathogenic member of the genus and most important human pathogen is *S. pyogenes*, which causes suppurative diseases such as tonsillitis and pharyngitis (with or without scarlet fever), sinusitis, and their complications; the skin infections cellulitis, erysipelas and impetigo and also puerperal sepsis. Rheumatic fever and acute glomerulonephritis are two very important non-suppurative sequelae to pharyngeal and skin infections. Lancefield group B streptococci (*S. agalactiae*) remain an important cause of bovine mastitis, and have emerged as causes of neonatal meningitis and septicaemia in humans.

2 α-Haemolytic (viridans) species: oral streptococci are of quite low pathogenicity but they may cause various opportunistic infections and are of major importance in dental plaque and caries, with *S. mutans* (which produces sticky dextrans and acids from dietary sugar) the chief offender. Dental bacteraemia may lead to colonization of damaged heart tissue, and oral streptococci are the commonest agents of subacute bacterial endocarditis.

3 *Streptococcus pneumoniae* has a wider host range than *S. pyogenes* and is the commonest and most important pathogen in the genus, but it is as a pathogen of humans that it is of particular importance, causing pneumonia, meningitis, septicaemia, otitis media, sinusitis and conjunctivitis.

4 Enterococci are hardy organisms resident in intestinal tracts and they are used as indicators of faecal pollution in water and foods; they are of low pathogenicity but *Enterococcus faecalis* and *E. avium* are opportunistic pathogens, and antibiotic resistance can make endocarditis and urinary tract infections by the former difficult to treat.

Lancefield group C and G streptococci can cause pharyngitis and other infections in humans, but organisms with these antigens are more prominent as veterinary pathogens, causing bovine mastitis and a variety of invasive infections of domestic and laboratory animals. *Streptococcus equi* is the causative agent of strangles in young horses, a disease characterized by abscesses of the pharyngeal region. Turning to the 'other streptococci' group, *S. suis* is responsible for meningitis, bronchopneumonia and septicaemia in young pigs, and the starch hydrolysing *S. bovis* is an important part of the normal rumen microflora, but is also associated with the alimentary disorders acidosis and bloat.

The lactococci, *S. thermophilus* and *Leuconostoc* species are of fundamental importance in the dairy industry as they are the chief components of starter cultures for acid and flavour development in fermented milk products such as butter, cheese and yoghurt. The many millions of tons that are produced each year represent about one-fifth of the world's fermented food and drink turnover. *Lactococcus lactis* strains produce the antibiotic nisin, which is active against Gram-positive organisms and has found application as a food preservative. Several staphylococci have a role in the flavour development of fermented sausages; *Staphylococcus carnosus,* in particular, is used in starter cultures for dry sausage.

Identification

The identification of the principal pathogenic staphylococci is quite straightforward — they are recognized by their growth on selective media and colonial and microscopic appearances, and confirmed by simple tests such as coagulase with a few easily determined key characters to separate the coagulase-negative species. None the less, the great clinical importance of these organisms and the need for rapid results have encouraged the commercial development of test kits. There are agglutination kits for detecting clumping factor and protein A, which are properties of the cell wall of *S. aureus*, and there are several kits which are based upon conventional biochemical tests and chromogenic enzyme substrates and yield identifications within 5–24 hours. In several cases additional tests such as coagulase, novobiocin sensitivity and anaerobic growth are recommended.

Apart from these, it should be noted that the early recognition of MRSA is most important and that much effort has been expended on the development of epidemiological typing systems for *S. aureus*. Bacteriophage typing is the most widely used method, but various chemotaxonomic and genetic approaches including determination of plasmid profiles by agarose gel electrophoresis, restriction endonuclease plasmid digest profiles, Southern blotting with DNA hybridization probes for chromosomal transposons, and [^{35}]S methionine protein electrophoresis have been used successfully.

The streptococci commonly pathogenic to humans are also easily identified, by colonial morphology, haemolysis, disk sensitivity tests, microscopy and a few other key characters. For the β-haemolytic species Lancefield grouping remains the best method, and commercial test kits based upon latex particle or protein A-rich staphylococcal cell agglutination are widely used, and backed up with a small number of routine biochemical or physiological tests. Monoclonal antibodies can be used to differentiate enterococci from other Group D streptococci, and to identify *S. mutans* and other oral species. For the pneumococcus, a different serological technique, the capsule swelling test, is used; microscopically observed capsules become more visible and appear to expand in the presence of the homologous antiserum or pooled antisera. The enterococci and oral streptococci are differentiated by routine physiological and biochemical tests, but because many characters are inconsistent within individual species, a best fit or spectrum analysis approach is adopted — that is to say the identification is based upon overall patterns of results over a relatively large number of characters rather than being limited to a few unreliable key characters. Selections of such tests are commercially available as kits which, like those for the staphylococci, contain conventional and chromogenic enzyme substrates. Another, novel, approach described for enterococci is based on activity patterns of bacteriolytic enzymes. The chemotaxonomic and molecular genetic methods used for classification as described above are also of value and, for example, systems have been developed for the identification of lactococci, enterococci and *Leuconostoc* species by colony hybridization with 23S rRNA-targeted oligonucleotide probes or by PCR assisted probing; however, for many laboratories such approaches are not yet suitable for routine use.

13 The mollicutes

It is safe to speculate that only a minute fraction of the numbers of mollicutes occurring in nature have been grown and characterized . . . it is quite possible that the mollicutes might be the most prevalent prokaryotes in existence, and the ultimate in biological diversity.

J.G. Tully, 1989, *The Mycoplasmas V*

Introduction

The mollicutes (previously known as the mycoplasmas) are small, free-living prokaryotes without cell walls and with unusually small genomes and low mol% G+C ranges; they also have unusual nutritional requirements and will pass through bacteriological filters. The story begins in Germany in 1693 with the recognition of contagious bovine pleuropneumonia, a fatal disease of cattle. Pasteur is credited with suggesting a specific ultramicroscopic agent as the cause, Nocard and Roux cultivated the agent *in vivo* in 1898, and in 1900 it was first grown on a solid medium, yielding characteristic, tiny, fried-egg-like colonies (Fig. 13.1). Despite its growth on cell-free media, its filterability resulted in a widespread belief that it was a virus. Similar agents were sought and found in a wide variety of animals (sheep and goats, 1923; dogs, 1934) and from other sources (sewage, 1936; soil, 1937), and the frequent contamination of tissue cultures by these pleuropneumonia-like organisms (PPLOs) was not the least of the reasons for their intensive investigation.

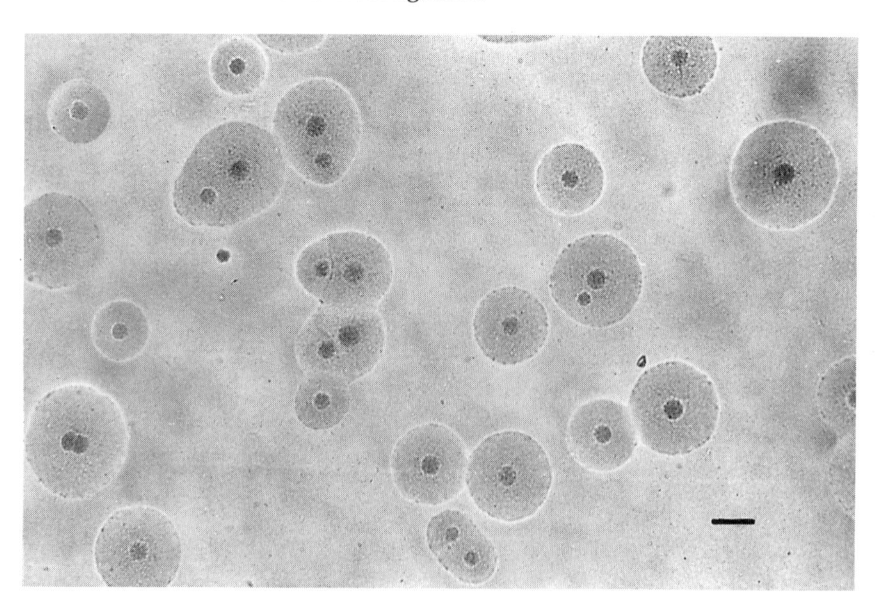

Fig. 13.1 Colonies of a strain of *Mycoplasma simbae*, isolated from the throat of a lion, grown on an agar medium for 2 days under aerobic conditions. Bar marker represents 100 μm. Reproduced with permission from A.C. Hill (1992) *International Journal of Systematic Bacteriology* **42**, 518–523.

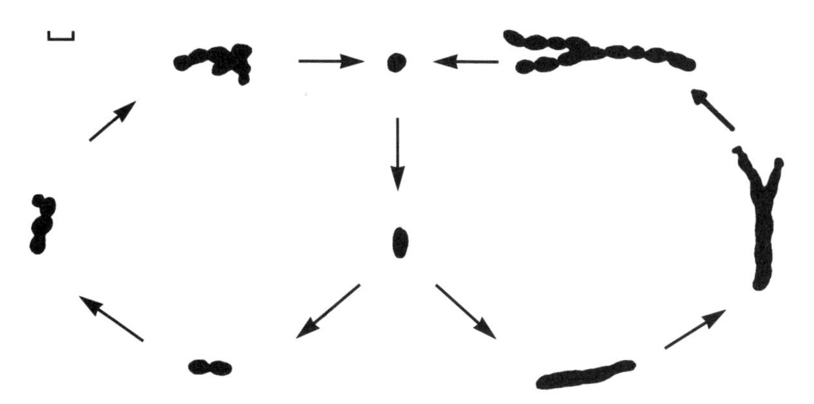

Fig. 13.2 Replication of a typical mollicute. Filamentous growth is normally associated with young cultures growing exponentially under optimum conditions; the multinucleate filaments later constrict between genomes and transform into chains of cocci which subsequently separate. Typical binary fission occurs when cytoplasmic division and genome replication are synchronized; budding may represent binary fission with unequal division of the cytoplasm. Bar marker represents 1.0 µm.

In 1929 the name *Mycoplasma* ('fungus form') was proposed for these organisms, on account of their filamentous nature (Fig. 13.2), and the first human isolate was reported in 1937. In 1935 however, Kleineberger described PPLOs growing, as she thought, symbiotically with the bacterium *Streptobacillus moniliformis* but in 1939 Dienes showed that these L-forms (L from Kleineberger's Lister Institute strain designation) were derived from the bacterium and could revert to the walled form of *S. moniliformis*. Kleineberger did not accept this explanation for 10 years, but one of her strains was indeed a mycoplasma.

The controversy that ensued was accompanied by the observation of L-forms in a wide variety of bacteria and it was shown that they could be induced by penicillin and other antibiotics that interfere with cell wall biosynthesis, bacteriostatic chemicals, bacteriophage and antibodies. There emerged two schools of thought on the relationship of PPLOs to other prokaryotes: one that they formed a true class of phylogenetically related organisms (some believing them to represent survivors of primitive cell types that preceded present-day bacteria), and the other that they were merely a heterogeneous collection of bacteria without cell walls. Genomic studies have revealed that the first part of the first of these hypotheses is correct.

By 1955 enough information had been gathered on the morphology, physiology, immunology and pathogenicity of the mycoplasmas to permit a tentative classification to be made and 15 species were assigned to the genus *Mycoplasma* in the family *Mycoplasmataceae*, order *Mycoplasmatales* in the seventh edition of *Bergey's Manual*. Research on these organisms increased greatly during the 1960s as their medical and veterinary importances became better appreciated, and their smallness and lack of cell walls attracted the attention of molecular biologists. In 1967 it was proposed that their diversity and ubiquity, taken along with their similar colonial and cellular morphologies and other characteristic features, warranted elevation to the status of class, with the name *Mollicutes* ('pliable skin'). Since then the class has expanded to contain three orders, four families, six genera, and over 130 species, and numerous mycoplasma-like organisms (MLOs) which await cultivation and better characterization have been isolated from diseased plants and other sources.

The class has also contracted as the monospecific genus *Thermoplasma* has to be excluded. *Thermoplasma acidophilum*, an inhabitant of heated coal tips, is a member of the *Mollicutes* by definition as it is a wall-deficient, self-replicating prokaryote; however, its special habitat, thermoacidophilic nature, and resistance to osmotic lysis raised doubts about a close relationship with the other parasitic mollicutes. Recent chemotaxonomic and genomic data indicate that it belongs to the archaea: the salient features are flagellar motility, simple nutrition, typical archaeal ether lipids, presence of cytochromes and quinones not found in mollicutes, absence of the eubacterial proton-translocating ATPase, high mol% G+C, presence of a DNA-associated histone-like protein, DNA-dependent RNA polymerases like those of archaea, and certain nucleotide sequences in 5S rRNA, 16S rRNA and tRNA. Its relationship to the other archaea was considered in Chapter 6.

Phenotypic classification

The mollicutes are easily distinguished from other prokaryotes by their size, morphology, colonial characters, susceptibility to osmotic lysis and ability to pass through 0.45 μm membrane filters. Separation from the L-forms is less straightforward, but the latter are less uniform in size and do not produce the characteristic mollicute filaments (Fig. 13.2), but have cell wall polymer precursors and penicillin binding proteins in their membranes.

The mollicutes are not easy to study; their genomes are believed to be the smallest of any free-living cell and to approach the minimal coding capacity for cell function. The limited biosynthetic capability that this indicates is reflected by a paucity of phenotypic characters for classification work, so they are not amenable to conventional numerical taxonomic methods. Also, some species show considerable intraspecific heterogeneity in DNA homology and restriction enzyme analyses, so that these methods may be of limited value. The division of the class is reliant upon relatively few features (Table 13.1). Data for one of these, genome size, are of doubtful value as recent determinations by pulsed-field gel electrophoresis give results at variance with those obtained by the conventional approach using DNA renaturation kinetics, and show that mollicute chromosomes span a continuous range of sizes. Another feature which may be of taxonomic value is the production, by some species, of specialized cell tips (terminal structures) which appear to be involved in their attachment to host cells and, possibly, their gliding motility.

Table 13.1 Divisions of the class *Mollicutes* and other prokaryotes lacking cell walls

Classification	Number of species	Genome size*	mol% G+C	Cholesterol needed	NADH oxidase location: Membrane	NADH oxidase location: Cytoplasm	Habitat	Special features
Order I *Mycoplasmatales*†								
Family I *Mycoplasmataceae*								
Genus I *Mycoplasma*	95	4–8.7	23–41	+	–	+	Animals, insects and plants	
Genus II *Ureaplasma*	5	6	27–30	+			Animals	Ureolytic
Family II *Spiroplasmataceae*								
Genus I *Spiroplasma*	14	6.2–12.3	25–31	+	–	+	Arthropods and plants	Helical filaments
Order II *Acholeplasmatales*								
Family I *Acholeplasmataceae*								
Genus I *Acholeplasma*	14	10–11	27–36	–	+	–	Animals, arthropods and plants	
Order III *Anaeroplasmatales*								
Family I *Anaeroplasmataceae*								
Genus I *Anaeroplasma*	4	11	29–33	+	?	?	Rumen	Anaerobic
Genus II *Asteroleplasma*	1	11.5	40	–	?	?	Rumen	Oxygen sensitive
Other groups								
Thermoplasma	1	8.4–10	46	–	+	–	Heated coal tips	Archaeal thermo-acidophile
Mycoplasma-like organisms (MLOs)	?	4.3–8	?	?	?	?	Plants with yellows diseases and insects	

* Expressed as 10^8 daltons.

† A recent taxonomic revision proposes that *Mycoplasma* and *Ureaplasma* be restricted to strains primarily associated with vertebrates and that a monophyletic cluster of arthropod-associated strains currently classified as *Mycoplasma*, *Acholeplasma* and *Spiroplasma* species be elevated to ordinal rank. The new order, *Entomoplasmatales*, comprises the families *Spiroplasmataceae* and *Entomoplasmataceae*, the latter containing the non-helical species in the new genera *Entomoplasma* and *Mesoplasma* (Tully *et al.*, 1993).

Mycoplasma

Many species of *Mycoplasma* are restricted to, or are mainly associated with, particular hosts and tissues but others are more widely distributed. Restriction endonuclease cleavage patterns for the human urogenital tract parasite *M. genitalium* and the avian respiratory tract pathogen *M. gallisepticum* support the idea that the more specialized species may have high genetic homogeneity. There are only 10 widely useful routine tests for separating the 95 species of *Mycoplasma*, (chief among these being the ability to metabolize glucose or arginine, or both, or neither) and much reliance is placed upon mol% G+C determinations and serological methods such as growth inhibition, metabolism inhibition and immunofluorescence for their differentiation. The rapid increase in numbers of species and the rather wide G+C range of over 17% led to debate about the maintenance of a single genus, and the absence of common antigens and some correlations between phenotypic tests and G+C gave some support to the idea of splitting the genus, but the Subcommittee on the Taxonomy of the *Mollicutes* wished to await further data. Whether or not the widely used limit of a 10% range in G+C for a genus is respected, the genus is a diverse one as DNA homology studies have not shown significant genomic relatedness between the species studied to date. Conversely, according to PAGE, DNA relatedness and serological studies, the subspecies of *M. mycoides* (*M. mycoides* subsp. *mycoides* and *M. mycoides* subsp. *capri*) are very closely related to each other and to unnamed strains isolated from cattle and goats, and the retention of these subspecies has been questioned.

Ureaplasma

The genus *Ureaplasma* was established to accommodate the urea hydrolysing T-mycoplasmas (T for tiny colonies) isolated from humans and animals. Initial classification represented a lumping approach, which contrasts with the splitting that has occurred in *Mycoplasma*, and two of the five species represent collections of strains from particular hosts. The human strains, classified as *U. urealyticum*, comprise at least 14 serogroups and the bovine strains, *U. diversum*, comprise three. These species show narrow ranges of G+C content but they are not genetically homogeneous, and DNA hybridizations (with human strains) and PAGE and two-dimensial PAGE analyses support the division of the human strains into two species and the bovine strains into three (which correlate with the serogroups). However, more DNA relatedness information on the bovine strains is needed and the Subcommittee felt that some correlation with pathogenic or other phenotypic, and diagnostically useful, features was desirable for the human strains. Serologically distinct ureaplasmas have been isolated from a wide variety of other animals including non-human primates, dogs, sheep and goats, but the majority have not been sufficiently characterized, or too few strains have been isolated, to allow species proposals. Exceptions are *U. gallorale*, a serologically homogeneous species from chickens, and

U. felinum and *U. cati* which represent two groups of feline strains which are distinct by serology and DNA relatedness.

Spiroplasma

Helical mollicutes were first discovered as plant pathogens and the agent of citrus stubborn disease, which was the first to be cultivated axenically, was described and named *S. citri* in 1973. Several other agents previously thought to be viruses, spirochaetes or pleomorphic mollicutes were soon shown to be spiroplasmas. Since then these organisms have been used as models for membrane research and have attracted particular interest on account of their colonization of plants and arthropod tissues, pathogenicity for plants and arthropods, maintenance of a helical shape without a rigid cell wall, and non-flagellar motility, and many new ones have been discovered. Many of these were found on plant surfaces, but it was shown that they could be transmitted by insects and most recent new strains have been isolated from arthropods.

This rapid expansion of the group made a diagnostically satisfactory classification scheme necessary and, in addition to serology (particularly observation of the deformation of helical morphology by antiserum), two-dimensional PAGE and genomic methods such as genome size, mol% G+C, and restriction enzyme analysis, cloning and sequencing of DNA have been widely used. Over 30 different spiroplasmas are now recognized; they have been classified into 23 groups, the first of which has eight subgroups, but only 14 species have been validly published because the remainder (bearing names such as rabbit tick spiroplasma and leafhopper spiroplasma) do not satisfy the Subcommittee's minimum requirements for mollicute species (DNA base composition, serology by growth inhibition and immunofluorescence, glucose fermentation, hydrolysis of arginine and urea, and other optional genotypic, serological and biochemical tests). Although these methods have helped to define the genus, there is still no satisfactory serological or molecular marker for it. This means that although the group is bound to expand in numbers and diversity as the vast reservoirs of arthropods and plants are screened, the recognition of non-helical members of the genus might be difficult.

Acholeplasma

A growth requirement for sterol was initially regarded as a key character for separating mycoplasmas from the true bacteria and so sterol-nonrequiring isolates from sewage and soil presented a taxonomic problem. This was exacerbated by the discovery that *M. granularum* (a species from pigs) did not have a requirement for sterol and, along with recommended test procedures, the family *Acholeplasmataceae* and genus *Acholeplasma* were established to accommodate them. A further 13 species, mainly from various vertebrate hosts but latterly also from plants and insects, have been described and several of the acholeplasmas have been shown to possess further features that distinguish them from other mollicutes. Examples of these features are: polyterpenol synthesis, the positional distributions of membrane fatty acids, the presence of superoxide dismutase and, as demonstrated by two-dimensional immunoelectrophoresis, a common antigen. Notwithstanding these, convenient and reliable biochemical or serological methods for the separation of the species are awaited. It is surprising therefore, that the close relationships between species that might be inferred from such difficulties of distinction are not borne out by DNA relatedness studies, as these have shown that the species are genotypically well separated — indeed, hybridizations between type strains of several species have shown little relatedness, sometimes none. Furthermore, genomic relatedness within the two species *A. laidlawii* and *A. axanthum* spans an unusually wide range, with 48–100% DNA homologies and markedly different restriction endonuclease digest electrophoresis patterns; such variation is attributable, perhaps, to the species' diversities of hosts and habitats.

Anaeroplasma

An anaerobic, bacteriolytic rumen organism described in 1966 was shown to be a mollicute in 1973 and, because it was mistakenly believed not to require sterol it was placed in *Acholeplasma* as *A. bactoclasticum*. Further anaerobic mollicutes including non-bacteriolytic strains were isolated soon afterwards and it became clear that cattle and sheep rumens support a variety of such organisms. Once the sterol requirement of *A. bactoclasticum* was demonstrated the genus *Anaeroplasma* was proposed to accommodate it and *A. abactoclasticum* which, as type species, contained the non-bacteriolytic strains. Other, sterol-nonrequiring strains had also been isolated and it was proposed that these be included in the genus; anaerobiosis being regarded as outweighing sterol-nonrequirement as a taxonomic character. However, this meant that the genus could not be included in the family *Mycoplasmataceae*, as the latter character was used for the exclusion of the acholeplasmas which, as we have seen, subsequently achieved family and ordinal status.

The Subcommittee on the Taxonomy of *Mycoplasmatales* agreed that such divergent strains should not be included in the same genus. When the type strains of the two species and other strains were subjected to agglutination, gel immunodiffusion precipitation and modified metabolism inhibition tests, four serogroups emerged with *A. bactoclasticum* comprising two of these. Studies of nucleic acid relationships showed that the anaerobes were not genomically related to the acholeplasmas, that the sterol-nonrequiring strains had much higher mol% G+C than the other anaerobic strains (Table 13.1), and that five groups representing distinct species existed. Surprisingly, two of these groups comprised one *A. bactoclasticum* serogroup. It was therefore proposed that four species of *Anaeroplasma* and one species of a new genus *Asteroleplasma* be established within a new family and order.

Mycoplasma-like organisms

Phenotypic studies of the MLOs have been hindered by failure to grow any of them *in vitro* and they are harder to work with than most plant viruses, so little is known of their interrelationships. They are associated with 'yellows' and other diseases in a wide range of plants worldwide and are observed within plant sieve-tube elements and the salivary glands of insect vectors. Until recently characterization was limited to host ranges, symptoms, vector specificities and microscopical information, all of which are very time consuming and difficult to obtain. Now nucleic acid data may be obtained by cloning fragments of MLO chromosomes and using them as probes, and polyclonal and monoclonal antibodies against MLO antigens can be used with ELISA and immunosorbent electron microscopy to detect these organisms in infected plants and insects.

Although MLOs have been reported in several-hundred plant species, many, if not most, of these will not represent different taxa, as several of the most studied agents have wide host ranges. A study of virescence MLOs (which produce greening of flowers and leaf-like sepals and petals) emphasized this point. Chromosomal fragments of western aster yellows cloned in *E. coli* were used to probe Southern blots of DNA from a wide variety of infected plants from different parts of the world; hybridization occurred with the majority of the virescence MLOs tested but not with decline and other MLOs or with spiroplasmas. Advances in serological and genomic methods, especially PCR, now promise a major step forward in our understanding of this large and important group of organisms.

Phylogenetic classification

As we have seen in other chapters, classifications derived from rRNA sequences are often in considerable disagreement with arrangements based upon large amounts of phenotypic data; since mollicute classification relies upon rather few phenotypic features, with help from genomic features whose usefulness at the species level and accuracy are in some doubt, it should come as no surprise that phylogenetic groupings currently show little congruence with the phenetic ones.

The first clues about the evolutionary origins of the mollicutes came in the 1960s from the recognition of similarities between the respiratory pathways of the fermentative mycoplasmas and the lactic acid bacteria: both groups ferment glucose by the Embden–Meyerhof pathway, accumulate lactate, have flavin-terminated respiratory systems, and do not synthesize haem enzymes. In the 1970s it was found that acholeplasmas possess lactate dehydrogenases activated by fructose-1,6-bisphosphate, a type of activation previously known only in streptococci, and from comparative enzyme immunology of aldolases it was subsequently concluded that the acholeplasmas evolved from Lancefield group D or N streptococci (now the enterococci and lactococci).

Later, 5S rRNA sequencing indicated that the acholeplasmas (represented by *A. laidlawii* and *A. modicum*) arose from the *Clostridium innocuum* group, presumably losing their cell walls when the genome was reduced, that this line then branched to yield two sterol requiring groups, the anaeroplasmas and the spiroplasmas, and that the latter then gave rise to the mycoplasma and ureaplasma lines with a further reduction in genome. A larger study based on selected 16S rRNA sequences of 46 mollicutes and several walled relatives confirmed the conclusions of the 5S rRNA study and revealed six distinct divisions (Fig. 13.3). These were defined by distance matrix analysis, sequence signatures (where members of each group share unique compositions at one or more positions in the RNA molecule), and in some cases, particularly the group containing *M. pneumoniae*, by unique features of the molecule's secondary structure.

The divisions of mollicutes and their relationships with walled bacteria were also supported by distributions and activities of pyrophosphate-dependent enzymes; in the majority of other bacteria ATP is the cofactor

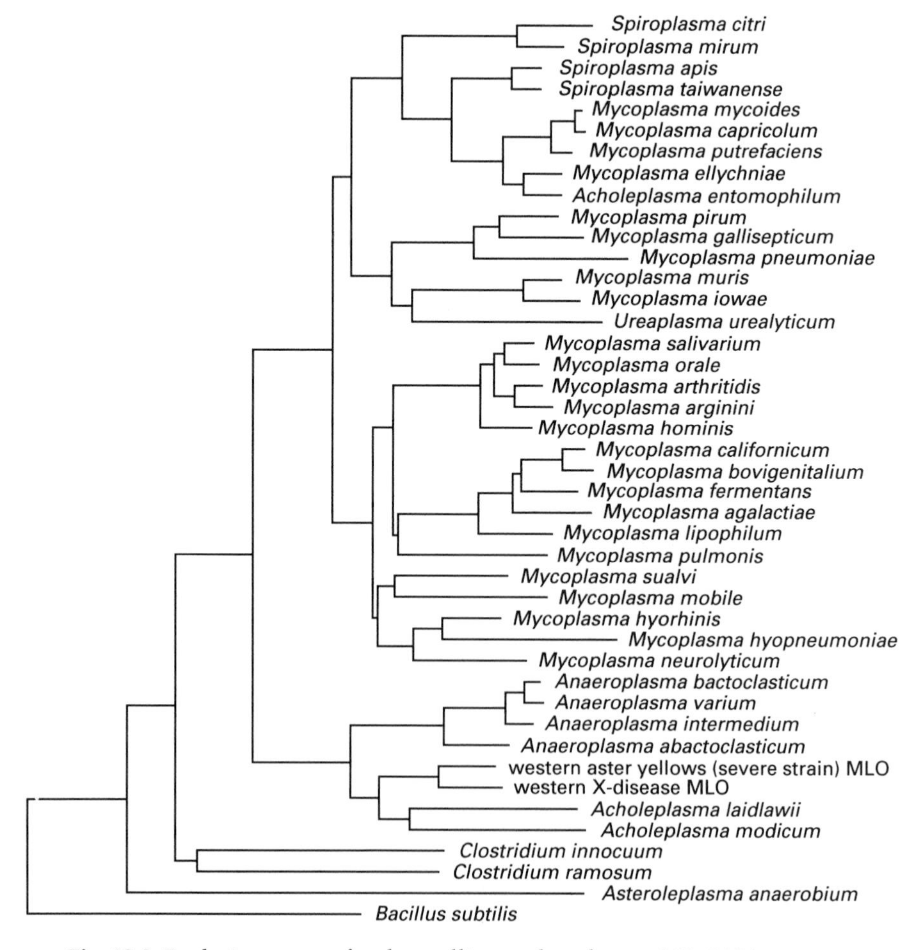

Fig. 13.3 Evolutionary tree for the mollicutes, based upon 16S rRNA sequence comparisons. Adapted from Weisberg *et al.* (1989) and Kuske & Kirkpatrick (1992).

for analogous enzymes. Other phylogenetic markers of potential value include rifampicin insensitivity (also characteristic of some clostridia), the use of UGA as a stop codon (rather than for tryptophan as it is in other bacteria), and analysis of the *tuf* gene which codes for a ubiquitous elongation protein. In higher organisms atypical phenotypes are regarded as evidence of a rapid pace of evolution. The mollicutes are certainly atypical compared with the rest of the eubacteria and further evidence for their rapid evolution is seen at the molecular level; some rRNA sequence positions that are usually highly conserved in other organisms are quite variable or absent in the mollicutes, and they also possess many unique oligonucleotides. Studies on another rapidly evolving organism, *Leuconostoc oenos*, have shown that such evidence may be found throughout the genome so that all the peculiarities of the phenotype are reflections of a rapid evolution.

With respect to the current classification it can be seen (Table 13.2) that the phylogenetic groups show little correlation with phenotypic characters. Two of the characters important at the order and family levels, sterol requirement and genome size, do not separate the phylogenetic groups; arginine hydrolysis is scattered widely, and only terminal structure and glucose fermentation show any alignment. As immunoblotting and other serological methods promise to improve the simplicity and accuracy of mollicute identification, their widespread adoption could reduce emphasis on other phenotypic characters for taxonomy, and therefore, diagnosis, and allow the class to be rearranged along phylogenetic lines.

Turning to the MLOs, the phylogenetic evidence is scanty to date, but tantalizing. The DNA region containing the gene for the 16S rRNA of an MLO pathogen of the evening primrose has been cloned and sequenced, and sequence comparisons with a small number of subsets suggested an evolutionary relationship with *Mycoplasma capricolum* but a closer relationship of the latter with *Bacillus subtilis*. This may imply evolutionary remoteness between the MLO and *Mycoplasma* or rapid evolution owing to a high mutation rate. There is evidence for both of these explanations: unlike most other bacteria the spacer region between the 16S rRNA and 23S rRNA genes contains a single $tRNA^{Ile}$ (isoleucine) gene, and this has been taken to support the idea of early divergence from the other mollicutes, while the absence of several of the 16S rRNA oligonucleotides that are highly conserved in other bacteria and the presence of unique sequences suggest a rapid evolutionary pace. Comparison of the occurrence of these unique oligonucleotides indicated an MLO signature resembling that of *A. laidlawii*. Furthermore, the 16S rRNA gene sequence of western aster yellows MLO implied a close relationship with another MLO causing similar symptoms, and more distant relationships with the western X-disease MLO, and *A. laidlawii* (Fig. 13.3); dot hybridizations and restriction endonuclease analyses indicate that various proliferation MLOs are distantly related to aster yellows strains. It is therefore noteworthy that in the larger 16S rRNA sequencing study (Table 13.2 & Fig. 13.3) *A. laidlawii* was recovered in the anaeroplasma group, which contains large-genome organisms believed to represent the early stages of mollicute evolution.

Table 13.2 Phylogenetic groupings of mollicutes and their phenotypic features*

Groups, subgroups and key species in subgroup	Number of species	Sterol requirement	Genome size	Glucose fermentation	Arginine hydrolysis	Terminal structure	Hosts
Pneumoniae							
Mycoplasma pneumoniae	3	+	4.9–7.1†‡	+	v§	+	Humans and birds
M. muris	3	+	6–8.7	v	v	–	Animals and birds
Ureaplasma urealyticum							
Hominis							
M. hominis	5	+	4–4.9	–	+	–	Humans and animals
M. lipophilum	5	+	4–7.7‡	–	v	–	Humans, cows and goats
M. pulmonis	3	+	6.3‡	+	v	+	Animals and fish
M. neurolyticum	3	+	7.1‡	+	–	–	Pigs and mice
Spiroplasma¶							
M. mycoides	6	v	4.8–10.6‡	+	–	–	Cows, goats, insects and plants
Acholeplasma florum							
Spiroplasma citri	2	+	10.6‡	+	+	–	Plants and arthropods
S. apis	2	+	?	+	v	–	Plants and arthropods
Anaeroplasma							
Acholeplasma laidlawii	2	–	10–11	+	–	–	Animals
Western aster yellows MLO	3	?	4.3–8**				Plants
Anaeroplasma abactoclasticum	3	+	11‡	+	–	+(?)	Pigs and cows
Asteroleplasma							
Asteroleplasma anaerobium	1	–	11.5	+	–	–	Pigs

* Adapted from Weisburg *et al.* (1989) and Kuske & Kirkpatrick (1992).

† Expressed as 10^8 daltons.

‡ Data not available for all members of subgroup.

§ Variable.

¶ See footnote † in Table 13.1.

** Range for MLOs.

Important species

Over a dozen species of mollicutes are carried by humans in the mouth and respiratory tract and in the genitourinary tract. The most important species and only unequivocal pathogen is *M. pneumoniae*. The peak incidence of clinically apparent infection caused by this organism (long known as primary atypical pneumonia to distinguish it from pneumococcal disease) is in 5–15 year olds and may account for up to 50% of the pneumonias in this group; much work has been devoted to developing a satisfactory vaccine. The species found in the mouth appear to be harmless commensals, but the two sexually transmitted, genitourinary species *M. hominis* and *U. urealyticum* have both been implicated in pelvic inflammatory disease and the latter appears to be associated with non-gonococcal urethritis.

Policies of slaughter and vaccination have eradicated contagious bovine pleuropneumonia from most countries but the disease, caused by *M. mycoides* subsp. *mycoides*, remains endemic and causes serious economic losses in many tropical areas. Other mollicute infections of domestic animals have mastitis, arthritis, conjunctivitis and pneumonia or other respiratory problems as their main symptoms, often together with systemic disease, and several are economically important worldwide. Cattle, sheep, goats, pigs, chickens, turkeys, dogs, cats and laboratory rats and mice are the principal animals affected and other species of undetermined pathogenicity are found in these and several other animals.

Mollicutes are also frequent contaminants of animal cell cultures, *A. laidlawii*, *M. arginini*, *M. hyorhinis* and *M. orale* being the commonest. They may have a variety of effects that can confuse investigations; these include mycoplasmal enzyme and toxic effects, pH changes, increases or decreases in viral yields, promotion or inhibition of interferon induction, inhibition of lymphocyte stimulation, interference with mutagenic assays, and chromosomal aberrations. Contamination is detected by a variety of cultural and non-cultural methods, but none of the many techniques described for eliminating it are entirely reliable and cell cultures are usually discarded if possible.

Awareness of the importance of mollicutes as plant pathogens has increased dramatically in recent years. Diseases such as corn stunt and citrus stubborn were long considered to be caused by viruses, but the discovery and culture of spiroplasmas from them in the early 1970s and the involvement of vectors heralded massive research interest in mollicutes of plants and arthropods. It has become clear that the latter carry a vast range of such organisms, particularly MLOs, which may cause disease in them or in plants to which they are transferred; although these have not yet been cultivated *in vitro*, genomic methods are facilitating their characterization and some 300 plant diseases have been described. Although these were commonly referred to as yellows diseases, MLOs cause other characteristic symptoms including virescence, phyllody (flowers become leafy), witches' brooms (proliferative growth), deformities, stunting and necrosis.

Identification

Animal mollicutes are identified using the characterization methods outlined above — the emphasis is on serological methods with high specificity such as colony inhibition on agar, metabolism inhibition and immunofluorescence. Of the plant and arthropod mollicutes, spiroplasmas are readily recognized by phase-contrast or dark-field microscopy and further identified by such serological methods as the spiroplasma deformation test and ELISA, but other organisms do not have morphologies distinguishable from their hosts' organelles and debris and all the MLOs remain uncultivable. Detection and identification of these organisms were formerly dependent on symptomatology and histological staining, but several recent technological advances promise to change that. Serological methods were previously only applicable to cultivable mollicutes because crude MLO antigen preparations yield nonspecific antisera that react with plant antigens, and adequate amounts of pure MLOs could not be obtained. The hybridoma technique is therefore very promising: essentially, mouse myeloma cells and antigen-activated B lymphocytes are fused, hybrid cells are selected for, and hybridomas are screened for production of the specific monoclonal antibodies required. The DNA of MLOs may be isolated by density gradient centrifugation (taking advantage of the low mol% G+C values of cultivable mollicutes and assuming that those of MLOs are also low), cloned or amplified by PCR, and used to make highly sensitive and specific identification probes.

14 The actinomycetes

The emerging natural taxonomy of the actinomycetes is not only of value in its own right but will also form an essential part of a developing microbial technology.

M. Goodfellow, 1985, *6th International Symposium on Actinomycete Biology*

Introduction

Until the late 1940s the 'ray fungi' were regarded as peculiar organisms of little practical importance and few studied them; they were widely regarded as fungi and the diseases they caused continued to be covered under medical mycology long after their bacterial nature was clear. Subsequent to Schatz & Waksman's discovery of streptomycin in 1943, however, there has been an explosion of interest in these organisms and their pharmaceutical, medical, veterinary, agricultural and ecological importances have become appreciated. The hyphal and mycelial growth of many species gives a primitive multicellularity and so they are also much studied for their own sakes.

The *Actinomycetales* is a vast order containing over 60 genera or groups at the generic level. Buchanan's original definition of the order in 1917 was on the basis of morphology, as members had in common a tendency to branch and form hyphae at some stage in their development, but although this feature is easily observed in some, with others considerable imagination is required. Molecular studies have largely confirmed the content of the order, several established groups such as *Arthrobacter*, *Corynebacterium* and *Micrococcus* have joined it, *Thermoactinomyces* has been the only important departure, and many new genera have been described.

Sequence analyses of 16S rRNA have shown that the actinomycetes belong to the Gram-positive, eubacterial evolutionary line. This phylum appears to comprise four subdivisions, only two of which are well characterized, and these two are roughly separable by their mol% G+C ranges. The typical Gram-positives such as *Bacillus*, *Clostridium*, *Lactobacillus*, *Streptococcus* and *Staphylococcus* are in the low G+C subdivision with (mostly) less than 50 mol% G+C (although *Bacillus*, with a range of 32–68%, confounds this observation spectacularly), whereas the high G+C subdivision with 55 mol% or more contains the *Actinomycetales*, and relatives such as the bifidobacteria and propionibacteria.

The actinomycetes show great morphological diversity; they range from coccoid organisms through others with coccus–rod cycles, non-branching rods, slightly branching rods, fragmenting hyphal forms, and permanent and highly differentiated branched mycelia with spore-bearing aerial hyphae, to motile spores and multilocular (many-compartmented) sporangia (Fig. 14.1). Morphology remains important for defining actinomycete groups at the genus and family level (Table 14.1), but chemotaxonomic approaches were introduced as early as the 1950s to complement this and to help in species differentiation, which relied upon a wide range of cultural and biochemical tests. Chemotaxonomic methods have had a greater impact on the classification of actinomycetes than on that of any other group and, along with numerical taxonomy and genomic data, they have permitted thorough overhaul of the order, as consideration of some of the more important and interesting groups will show.

Cell wall analysis for actinomycete classification was pioneered by Cummins & Harris with publications in 1956; since then eight types have been described (Table 14.2) and these have allowed the recognition of groups of genera at the family level or fit well with the families and less formal groupings of genera that have been recognized on the basis of other characters. For genera with cell wall chemotypes II–IV, which contain *meso*-DAP, whole-organism sugar patterns (Table 14.3) are also of value (Table 14.1). Peptidoglycan types (see Chapter 2), which are recognized on

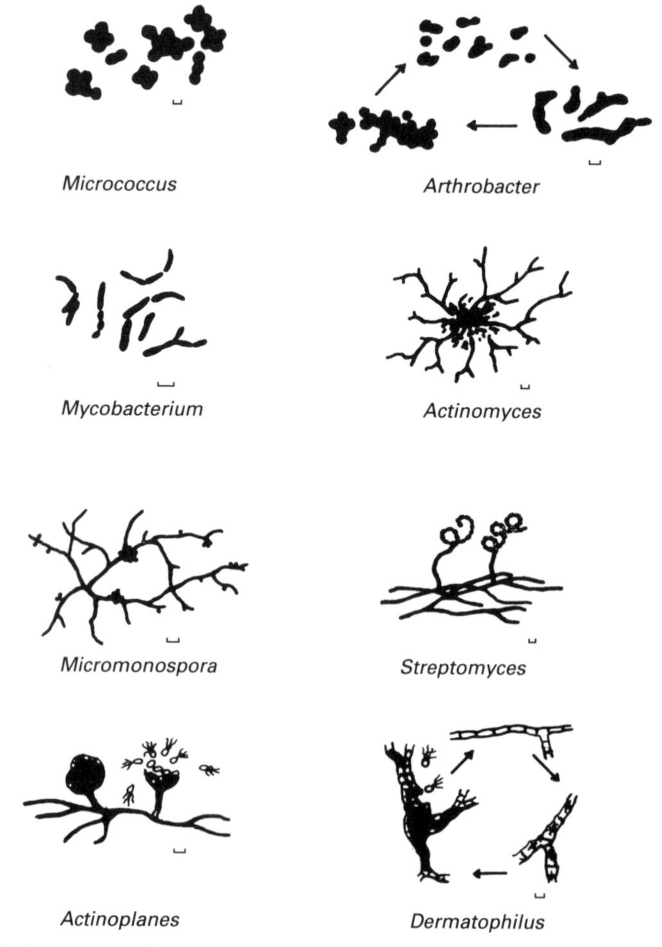

Fig. 14.1 Some examples to illustrate the great morphological diversity of the actinomycetes. Bar markers represent 1.0 μm.

Table 14.1 Suprageneric groups of actinomycetes arranged according to phylogeny

Family or genus, type genus and example genera	Number of genera in family	Typical morphology	Wall chemotype	Sugar pattern	Phospholipid type
Cellulomonadaceae	4	Irregular rods — branching fragmenting hyphae	VI VIII	–	PV
Cellulomonas					
Oerskovia					
"Micrococcaceae"	6	Cocci, coccus–rod cycle, rods	VI	–	PI
Micrococcus					
Arthrobacter					
Dermatophilaceae	1	Multilocular septate mycelium	III	B	PI
Dermatophilus					
Microbacteriaceae	5	Cocci — irregular rods — filaments	VI VII VIII	–	PI
Microbacterium					
Agromyces					
Brevibacteriaceae	1	Coccus–rod cycle	III	C	PI
Brevibacterium					
Actinomycetaceae	2	Cocci — irregular rods — branching	V VI	–	PII
Actinomyces					
Micromonosporaceae	4	Branched, non-fragmenting hyphae, aerial hyphae scanty or absent	II	D	PII
Micromonospora					
Actinoplanes					
Nocardiaceae	4		IV	A	PII
Nocardia		Unbranched — branched fragmenting substrate and aerial hyphae			
Gordona		Rods and cocci			
Rhodococcus		Cocci — rods — branched mycelium			
Tsukamurella		Rods and coccoid rods			
Mycobacteriaceae	1	Rods, occasionally branched, acid-fast	IV	A	PII
Mycobacterium					
Corynebacteriaceae	1	Irregular, clubbed rods	IV	A	PI
Corynebacterium					
Pseudonocardiaceae	7	Substrate and aerial mycelium bearing spores	IV	A	PII and PIII
Pseudonocardia					
Amycolata					
"Thermomonosporaceae"	5	Branched, with sporebearing aerial hyphae; extensive variation	III	B and C	PI PII and PIV
Thermomonospora		Single spores often in clusters		C	PIV
Actinomadura		Short chains of arthrospores		B	PI

continued on p. 214

Table 14.1 *continued*

Family or genus, type genus and example genera	Number of genera in family	Typical morphology	Wall chemotype	Sugar pattern	Phospholipid type
Streptosporangiaceae	6	Stable substrate and aerial mycelium	III	B and C	PIV
Streptosporangium					
Microtetraspora					
Streptomycetaceae	(2)	Chains of spores on aerial hyphae	I	–	PII
Streptomyces					
(*Streptoverticillium*)					
Frankiaceae	2	Multilocular sporangia	III		
Frankia				B C and E	PI
Geodermatophilus				C	PII
Nocardioidaceae	2	Branching, fragmenting mycelium with aerial hyphae	I	–	PI
Nocardioides					
Other genera					
Actinobispora		Pairs of spores on substrate	IV	*	PIV
Glycomyces		Short spore chains on aerial hyphae	II	D	PI
Intrasporangium		Fragmenting, branching mycelium	I	–	PI
Kibdelosporangium		Aerial sporangium-like structures	IV	A	PII
Kineosporia		Substrate mycelium	I	–	PIII
Nocardiopsis		Fragmenting hyphae	III	C	PIII
Sporichthya		Aerial mycelium, motile spores	I	–	ND

* Novel sugar pattern comprising arabinose, galactose and xylose.
ND, not determined.

Table 14.2 Cell wall chemotypes

	Chemotype							
Major constituent	I	II	III	IV	V	VI	VII	VIII
L-Diaminopimelic acid	+							
meso-Diaminopimelic acid		+	+	+				
Diaminobutyric acid							+	
Aspartic acid						v		
Glycine	+	+					+	
Lysine					+		v	
Ornithine					+			+
Arabinose				+				
Galactose				+		v		

v, variable amounts.

Table 14.3 Whole-organism sugar patterns of actinomycetes containing *meso*-diaminopimelic acid (chemotypes II–IV)

	Sugars				
Pattern	Arabinose	Fucose	Galactose	Madurose*	Xylose
A	+		+		
B				+	
C	No diagnostic sugars				
D	+				+
E		+			

* Madurose is 3-0-methyl-D-galactose.

Table 14.4 Characteristic phospholipid types of actinomycetes

	Type				
Phospholipid	PI	PII	PIII	PIV	PV
Phosphatidylinositol	+	+	+	+	+
Phosphatidylglycerol	v		v		+
Phosphatidylethanolamine		+	v	v	
Phosphatidylcholine			+		
Phosphatidylmethylethanolamine			v	v	
Phospholipids containing glucosamine				+	+

v, variable.

the basis of mode of cross-linking and the amino acid composition and sequences of cross-links, also correlate well with generic groupings, as do fatty acid patterns, phospholipid types (Table 14.4) and predominant menaquinone composition. Additionally, wall chemotype IV organisms currently fall into two groups, a division that is supported by phylogenetic studies, as the cell walls of members of the genera *Corynebacterium*, *Gordona*, *Mycobacterium*, *Nocardia*, *Rhodococcus* and *Tsukamurella* contain characteristic waxy esters of **mycolic acids**, whereas members of the family *Pseudonocardiaceae* lack mycolic acids.

Streptomycetaceae

The family *Streptomycetaceae* is now restricted to *Streptomyces*, which is one of the most important genera of the actinomycetes. The majority of actinomycetes isolated from the soil and most other habitats, using conventional isolation procedures, are *Streptomyces* species and many produce useful metabolites such as antibiotics. Consequently they have been much studied and numbers of species far exceed those of any other actinomycete genus or, indeed, family.

In 1976 Gordon pointed out that the classification of actinomycetes, on the basis of chemotaxonomic tests, electron microscopy and other time-consuming determinations on a few strains — but without correlating routine tests — was becoming as unworkable as that of *Salmonella,* where every serological difference meant a new species name. This problem was particularly true of *Streptomyces,* a genus proposed by Waksman & Henrici in 1943 to distinguish these aerobic sporeformers from *Actinomyces,* whose members are facultatively anaerobic. From the 1940s onwards the great interest in streptomycetes as producers of antibiotics resulted in widespread environmental isolation and screening programmes and metabolites were more readily regarded as novel if their producing strains had been described as new species; the patenting laws thus encouraged proposals of new species. As numbers of species increased, so did the difficulties of proving the validity of new taxa since satisfactory reference strains were not generally available; species were often proposed on the basis of a few trivial morphological and cultural differences and synonyms proliferated.

Thus, numbers of *Streptomyces* species rose from around 40 before 1940 to well over 1000 by 1957 and to about 3000 by the late 1960s, many of these cited only in patent literature. Several attempts were made to delimit species and species groups on the basis of spore shape, colour and chaining, substrate pigmentation and melanin production, but different authors gave different weights to these characters and so there was no agreed classification. International cooperative studies between the late 1950s and early 1960s evaluated the various taxonomic characters and selected the most reliable. In 1964 the International *Streptomyces* Project was set up to provide reliable descriptions of authentic cultures, based upon rigorously standardized observations of the characters listed above and carbon source utilization; standard cultures were deposited in culture collections as valuable reference sources, opening the way for comprehensive numerical taxonomic studies.

The earliest numerical taxonomies had, inexplicably, little effect on streptomycete classification and the 1974 edition of *Bergey's Manual* listed 463 recognized species, grouped according to colour, and over 80 others. A major numerical taxonomy published in 1983 allowed the recognition of many species synonyms, and led to the reduction of the genera *Actinopycnidium, Actinosporangium, Chainia, Elytrosporangium, Kitasatoa* and *Microellobospora* to synonyms of *Streptomyces,* and to the construction of a probability matrix for identification. This allowed the genus to be pruned to just 142 species for the 1989 edition of *Bergey's Manual,* but subsequently over 40 new species and numerous subspecies have been described (many of these, unfortunately, on the basis of single or very few strains), and over 40 revived names and new combinations have been proposed. Groups circumscribed in the numerical taxonomic study were later supported by data derived from DNA hybridization, protein electrophoresis, phage host range studies and serological analyses. A later numerical taxonomy based upon miniaturized physiological tests showed general agreement with the earlier study, indicated that the genus was still overspeciated, and supported the suggestion that *Streptoverticillium* be reduced to a synonym of *Streptomyces*.

Nucleic acid studies were initially hampered by the lack of adequate phenetic classifications from which to work. The later DNA relatedness data supported the groupings revealed by numerical taxonomy but the wide ranges of variation observed made consistent correlations with phenetic similarity and, therefore, diagnostic tests, difficult. Ribosomal RNA studies have been rather few given the importance of the group. Limited DNA–rRNA hybridizations and rRNA oligonucleotide catalogues support the recognition of the family *Streptomycetaceae* (Fig. 14.2), comprising the genera *Streptomyces* and *Streptoverticillium*, and the first of these approaches gave some support to the continued separation of the two genera. However, although clusters revealed by oligonucleotide cataloguing

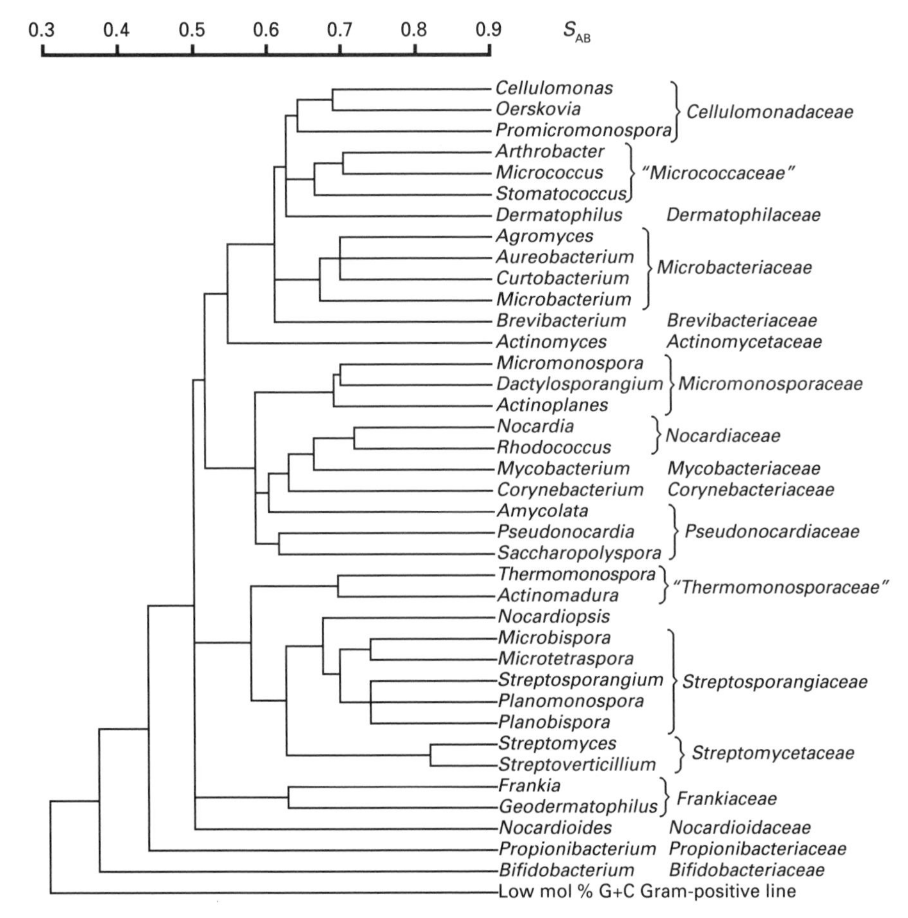

Fig. 14.2 Evolutionary relationships of actinomycetes based upon 16S rRNA partial sequencing.

showed some correlation with those from phenetic studies, the overall clustering did not separate the genera and it was proposed that both should be accommodated in an emended *Streptomyces*.

The taxonomy of *Streptoverticillium* lagged behind that of *Streptomyces* but suffered from similar problems. Its members were distinguished by their hyphae bearing whorls of short, spore-producing, branches (verticils) which give a barbed wire appearance; although the genus was morphologically quite distinctive, taxonomic work was hampered by a profusion of synonyms and inadequate numbers of strains for study. Further emendation of *Streptomyces* occurred when *Kitasatosporia* was reduced to synonymy with it. *Kitasatosporia* was established to include streptomycete-like strains with *meso*-DAP in the cell walls of substrate hyphae and L-DAP in the aerial hyphae, along with glycine and galactose, whereas the typical *Streptomyces* cell wall contains L-DAP with glycine (chemotype I); further cell wall analyses, as new species were described, confused the picture as the amounts of the DAP isomers varied ac cording to species, growth conditions and sporulation. Given this variability, it was considered that the phenotypic similarities, and strong hybridization of *Kitasatosporia* 16S rRNA with a *Streptomyces* probe, supported unification of the genera.

Strains of *Kineosporia* and *Sporichthya* are very rare, and both genera are of uncertain affiliation. They are usually considered with *Streptomyces* on chemotaxonomic grounds (Table 14.1).

Mycolic acid-containing actinomycetes

The term nocardioform, meaning actinomycetes that form a transient mycelium which breaks up into rod-shaped or coccoid elements, was long in use. Based as it was on morphology, the term brought together, informally, a large number of superficially similar but often not closely related organisms, individual strains of which may not have had true nocardial morphologies. Such organisms are mainly found in two groups, one comprising the three families *Corynebacteriaceae*, *Mycobacteriaceae* and *Nocardiaceae*, and the other the single family *Pseudonocardiaceae* (Table 14.1); while the two groups have similar wall chemotypes, whole-organism sugar patterns and peptidoglycan types, and have some commonality in phospholipid types, their menaquinone patterns show appreciable differences, they represent different evolutionary lines, and they are clearly separable by the fact that only members of the former produce mycolic acids.

The mycolic acid-containing actinomycetes represent a spectrum of nocardioform character ranging from *Gordona* with rods and cocci, and *Tsukamurella* with rods and coccoid rods, through the irregular rods of *Corynebacterium*, and occasionally branching rods of *Mycobacterium*, to *Rhodococcus* whose mycelium fragments into irregular rods and cocci, and *Nocardia* which produces a fragmenting mycelium and usually some aerial hyphae. *Corynebacterium*, *Mycobacterium* and *Nocardia* contain species of major medical interest.

Limited rRNA partial sequence analyses supported arrangements based upon chemotaxonomic data and showed that *Nocardia* and *Rhodococcus* are closely related and that they in turn are close to *Mycobacterium* (Fig. 14.2). The position of *Corynebacterium* was less clear; in one study it was quite close to *Nocardia* and *Rhodococcus*, but subsequent reports from the same laboratory indicated that the mycolateless actinoplanetes (now *Micromonosporaceae*) and even members of *Pseudonocardiaceae* lay closer to these genera and *Mycobacterium*. However, as the above-noted chemotaxonomic properties shared by most of the mycolic acid-containing actinomycetes are unlikely to represent convergent evolution, the closer relationship with *Micromonosporaceae* was in doubt. It is now conceded that the exclusion of *Corynebacterium* from the other genera of mycolate producers was an error owing to a quirk of evolutionary rates and the statistical methods used, and this emphasizes the importance of evaluating phylogenies in the light of phenetic data.

Nocardiaceae

As might be expected, the genus *Nocardia* was, for many years, a dumping-ground for nocardioform organisms as they were assigned on morphology alone, but in the light of numerical taxonomies and chemotaxonomic data many species have been reassigned or transferred to new genera which include *Amycolata*, *Amycolatopsis* and *Saccharopolyspora* (these three are now in the family *Pseudonocardiaceae*), *Actinomadura*, *Gordona*, *Nocardiopsis*, *Oerskovia*, *Rhodococcus* and *Rothia*, so that successive editions of *Bergey's Manual* recognize 45, 31 and 11 species respectively.

The genus *Nocardia* can now be defined on the basis of peptidoglycan composition, fatty acid pattern, 40–60 carbon mycolic acids, menaquinone profile and the other chemotaxonomic characters shown in Table 14.1. The main problem remaining was a nomenclatural one. Nocard's original 1888 isolate from a case of bovine farcy was named *Nocardia farcinica* in 1889 and the Judicial Commission made it type species in 1954. It later transpired that the ostensibly identical cultures of this organism in the ATCC and NCTC represented quite different strains, the latter belonging to the mycobacteria on the bases of mycolic acids, immunodiffusion, mol% G+C, numerical taxonomy and phage sensitivity. In view of its uncertain status *N. farcinica* was replaced as type species by *N. asteroides* in 1985.

The group of organisms known as "*Mycobacterium rhodochrous*" or the 'rhodochrous complex' caused taxonomic problems for many years. For Gordon & Mihm in 1957, it was an assemblage of organisms with a wide range of morphological diversity which, whilst showing similarities to *Mycobacterium* and *Nocardia* could not be assigned to either with confidence. Biochemical and physiological tests supported the inclusion of further strains from some 11 genera, principally *Mycobacterium*, *Nocardia*, *Proactinomyces* and *Corynebacterium* (Gordon also correctly predicting that further strains might be found in *Arthrobacter* and

Brevibacterium), and the group was regarded as a generic intermediate. In the 1960s and 1970s the group awaited satisfactory delineation and a decision on whether it should be given generic status, or accommodated in an existing genus, and if so which from a choice of *Arthrobacter*, *Brevibacterium*, *Corynebacterium*, *Mycobacterium* and *Nocardia*.

Numerical taxonomies showed it to be a heterogeneous group, often with two or three distinct clusters emerging, closely related to *Nocardia* but of equivalent rank to it and to *Mycobacterium*; in the early 1970s several strains were assigned to the new genus *Gordona*. Resolution followed polyphasic taxonomic studies; members of the group are chemotaxonomically very similar to *Nocardia* but their mycolic acids have somewhat shorter chain lengths, which can be differentiated by chromatography, and they are also separable serologically, by immunodiffusion tests in particular, and by antibiotic resistance tests. The genus *Rhodococcus* was consequently revived to accommodate these organisms (Fig. 14.2), but it remained a heterogeneous taxon comprising two clusters separable by chemical and serological data and mol% G+C. Reverse transcriptase sequencing of 16S rRNA revealed that *Rhodococcus* was phylogenetically heterogeneous and so the genus *Gordona* was revived to hold the higher mol% G+C cluster, strains of which have longer chain mycolic acids and menaquinones. At about the same time the species *Corynebacterium paurometabolum* and *Rhodococcus aurantiacus* were reduced to a single species in the new genus *Tsukamurella*. With these revisions *Rhodococcus* was left as a homogeneous taxon.

Mycobacteriaceae

Initially the classification of the mycobacteria was overshadowed by the medical and veterinary importance of *Mycobacterium tuberculosis* and *M. bovis*, all other cultivable strains being referred to as atypical or anonymous mycobacteria and only worthy of note as sources of confusion in the clinical laboratory. The first attempt at systematic classification was made by Runyon in the 1950s. His classification was manifestly artificial, but it was workable; the *M. tuberculosis* complex was left as it stood, little could be said about the non-cultivable species such as *M. leprae*, and the atypical mycobacteria were divided into four groups on grounds of growth rate and pigmentation. The groups are indicated in Table 14.5. Three contain slow growers, which require a week or more to produce visible colonies: group I strains produce yellow pigment when exposed to light (photochromogens), whereas those of group II produce yellow to red pigments regardless of light exposure (scotochromogens from the Greek *skotos*, darkness) and group III strains do not produce pigment (non-photochromogens). Group IV strains are rapid growers, producing visible colonies within 7 days and were described as non-pigmented but some rapid growing scotochromogens were subsequently described.

The International Working Group on Mycobacterial Taxonomy was established in the late 1960s and several cooperative studies on test

Table 14.5 Progress in the classification of the mycobacteria according to various schemes*

Runyon's scheme	Numerical taxonomy	Phylogenetic studies†
IV RAPID GROWERS Non-pigmented	RAPID GROWERS	RAPID GROWERS
M. fortuitum	*M. fortuitum* complex *M. chelonei*, *M. senegalense*	*M. fortuitum* *M. senegalense* *M. chelonei*
M. smegmatis	{*M. smegmatis*, *M. phlei*	*M. flavescens* *M. smegmatis* *M. phlei*
	M. flavescens	
		SLOW GROWERS
		M. simiae
		M. terrae *M. nonchromogenicum*
		M. xenopi
SLOW GROWERS		
II SCOTOCHROMOGENS	SLOW GROWERS	PATHOGENS
M. gordonae *M. scrofulaceum* *M. flavescens*	*M. gordonae* complex (*M. scrofulaceum*)	*M. gordonae*
M. TUBERCULOSIS group	*M. tuberculosis* *M. bovis*, *M. microti*	*M. bovis* *M. marinum*
		M. scrofulaceum
I PHOTOCHROMOGENS		
M. kansasii *M. marinum* *M. simiae*	*M. kansasii* complex *M. gastri*	*M. kansasii*/*M. gastri*
		M. szulgai *M. malmoense*
III NON-PHOTOCHROMOGENS		
M. avium *M. intracellulare*	*M. avium* complex *M. intracellulare* (*M. scrofulaceum*) *M. xenopi*	*M. intracellulare* *M. avium* *M. paratuberculosis*
	M. terrae *M. nonchromogenicum*	
NON-CULTIVABLE	Affiliation unknown	
M. leprae	*M. leprae*	*M. leprae*

* Bracketing indicates specific relationship.

† Indentation indicates evolutionary distance from putative common ancestor (see Fig. 14.3).

evaluation and numerical taxonomy were undertaken. The Runyon classification of the genus became supplanted by a more natural classification supported by comprehensively described species and reliable identification tests, but the genus still falls readily into the two groups of slow and rapid growers. (A proposal for the latter to be placed in the new, cumbersomely named genus *Mycomycobacterium* generated little enthusiasm and lipid analyses, immunodiffusion, immunoelectrophoresis, bacteriophage typing and DNA relatedness do not support it.)

Several species complexes emerged, as can be seen from Table 14.5, and much effort continued to be expended on trying to elucidate their taxonomic structures, especially in the case of the MAIS (*M. avium–M. intracellulare–M. scrofulaceum*) complex and its relatives such as the wood pigeon mycobacteria and *M. paratuberculosis* as they include the most clinically significant of the atypical mycobacteria. The last-named species is an obligate pathogen whose growth *in vitro* requires addition of a mycobactin, but members of the MAIS complex may also show some mycobactin dependence. Mycobactins are a unique family of iron-binding hydroxamate compounds (siderophores) that are produced by the mycobacteria and relatives and which may be of chemotaxonomic value.

Although serology was the most reliable approach to distinction within the complex, and correlated well with lipid patterns, cross-reactions confused the picture: not all MAIS strains reacted with the available sera, and some autoagglutinated. Subsequently, the relationships within the group have been elucidated by a variety of methods including DNA homology, numerical taxonomy and DNA restriction patterns. The problems of interpreting the many bands of restriction patterns have been reduced by restriction fragment length polymorphism analysis and field inversion gel electrophoresis; in the former hybridization with specific DNA probes yields a simplified pattern, and in the latter large fragments are separated and analysed with endonucleases that have infrequently occurring cleavage sites. These studies have led to the conclusions that *M. scrofulaceum* is not a close relative of other members of the group and that *M. intracellulare* is a distinct species. It was also proposed that *M. paratuberculosis* and the wood pigeon mycobacteria be considered subspecies of *M. avium*.

These developments are borne out by comparative 16S rRNA sequencing using the reverse transcriptase method and, most important given the high similarities involved, the more accurate PCR method. Results for limited numbers of strains indicate that the mycobacteria are a distinct and closely related group with sequence similarities of greater than 94%. The slow growers appear to have evolved from the fast growers' ancestral line and form a distinct line of descent within which the overt pathogens form a very tight cluster (Fig. 14.3 & Table 14.5). Given the difficulty of characterizing the non-cultivable *M. leprae*, and the consequent uncertainty of its affiliation, 16S rRNA sequence data have been particularly valuable in showing an evolutionary root for this organism within the pathogenic mycobacterial group.

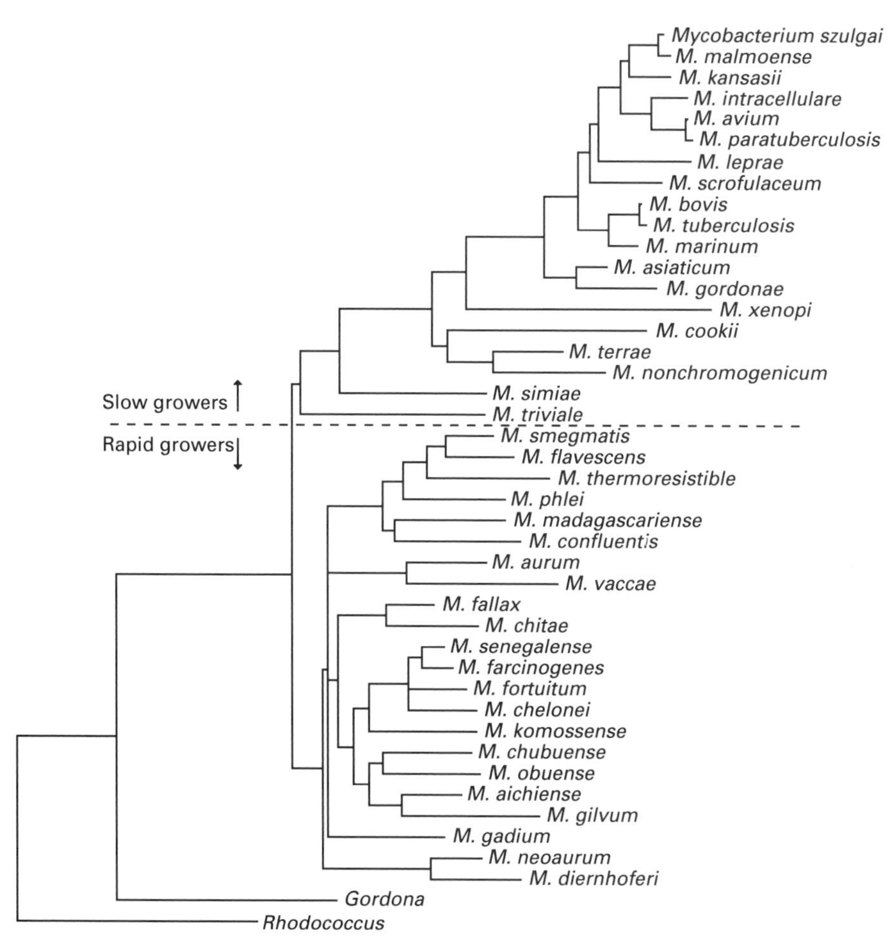

Fig. 14.3 A phylogenetic tree for the genus *Mycobacterium* based upon 16S rRNA sequence comparisons. Adapted from Pitulle *et al.* (1992).

Corynebacteriaceae

The story of *Corynebacterium* (meaning club-shaped rodlet) is not unlike that of *Nocardia*. The genus was created for *C. diphtheriae* and other pathogens but was subsequently used to accommodate an assortment of similar looking organisms known as diphtheroids and several other Gram-positive rods whose irregular shape and staining were covered by the term coryneform. A numerical taxonomy of coryneforms in the late 1960s included the names *Arthrobacter, Brevibacterium, Cellulomonas, Corynebacterium, Erysipelothrix, Jensenia, Kurthia, Listeria, Microbacterium, Mycobacterium* and *Nocardia*. To this list might be added organisms such as *Actinomyces, Arachnia, Bifidobacterium* and *Rothia* which have coryneform stages in their life cycles.

The heterogeneity of the group was obvious, and confirmed by numerical taxonomy, chemotaxonomy, mol% G+C and DNA relatedness, but there were differences of opinion as to where the boundary of the group should lie, and in the eighth edition of *Bergey's Manual* the authors spoke of the 'Coryneform Group of Bacteria'. This they divided artificially, but with good reason, into the human and animal parasites and pathogens, the plant pathogens and the non-pathogens. This division was not as artificial as it first appears: members of the first group produce mycolic acids and have similar cell wall sugars and wall chemotype IV, suggesting a relationship with the mycobacteria and nocardiae, whereas the plant pathogens lack mycolates, have quite different cell wall chemistries and show a somewhat higher mol%G+C range which imply that they belong to a different genus. The non-pathogen group contained a miscellaneous collection of organisms, many of which were poorly described.

Later work has fully supported this division. *Corynebacterium* is now defined by chemotaxonomic characters and is limited to the animal parasites and several adequately described non-pathogens in the mol%G+C range 51–63, which have mycolic acids of chain length between 22 and 36 carbons, arabinogalactan and *meso*-DAP in the cell wall, predominantly straight chain saturated and monounsaturated fatty acids, and menaquinones with two of the eight or nine isoprene units hydrogenated.

Chemotaxonomic studies showed the plant pathogen group to be heterogeneous and the 14 species are now accommodated in several genera: seven in *Clavibacter*, four in *Curtobacterium*, one in *Arthrobacter*, one in *Rhodococcus* and one in *Erwinia*. Of the remaining species, several have been transferred to other genera including *Arthrobacter*, *Rhodococcus* and *Tsukamurella* while others await fuller characterization. With improved definitions for several actinomycete genera the traffic has, not surprisingly, been two-way with the species *Arthrobacter variabilis* and *Brevibacterium ammoniagenes* being transferred to *Corynebacterium*.

Phylogenetic data on corynebacteria are sparse and, as mentioned earlier, two 16S rRNA analyses gave equivocal results. A 5S rRNA sequencing study placed species of the strictly defined genus close to members of the *Brevibacteriaceae*, *Microbacteriaceae* and "*Micrococcaceae*", and a single *Streptomyces* strain, but the absence of representatives of other mycolate-producing genera limited the conclusions that could be drawn.

Frankiaceae

This was originally proposed as a one-genus family, the genus accommodating symbiotic, nitrogen-fixing actinomycetes inhabiting root nodules (actinorhizae) of (in contrast with the rhizobiae) non-leguminous, woody plants and as such they have attracted much interest. Since early attempts to isolate *Frankia* strains and grow them *in vitro* were unsuccessful, species were defined on the basis of host plant cross-inoculation groups. As pure cultures became available from the late 1970s it was found that at least two of these host specificity groups, *Alnus* (alder) and *Elaeagnaceae* (wild olive family) could be recognized and separated by immunodiffusion, lectin-binding patterns, DNA relatedness, PAGE and gas chromatography of whole-organism sugars, but since compatibility groups were not always clear cut and did not correlate well with the conventional biochemical tests used elsewhere in actinomycete systematics, the 1989 edition of

Bergey's Manual gave no species descriptions and it was proposed that strains should be designated by letters and numbers according to the isolating laboratory. That edition included *Dermatophilus* and *Geodermatophilus* in the same group 'Actinomycetes with Multilocular Sporangia' for convenience, and the three genera do show some chemotaxonomic similarities (Table 14.1), but since then the taxonomy of these organisms at all levels has been radically altered by nucleic acid studies.

Partial sequencing of 16S rRNAs has shown that *Frankia* and *Geodermatophilus* are indeed related, and it has been proposed that the latter belongs within *Frankiaceae*, but that *Dermatophilus*, whose members are skin parasites, becomes the sole member of the family *Dermatophilaceae* (Fig. 14.2). With a large number of strains from a wide range of sources, DNA relatedness studies revealed at least nine species among three compatibility groups: three in *Alnus*, five in *Elaeagnaceae* and one in *Casuarina* (she-oak), and these groupings are supported by phylogenetic analysis using amplified 16S rDNA sequences. As *Frankia* or *Frankia*-like organisms have been reported in about 200 plant species representing 24 genera, and as morphologically similar but non-nodulating, free-living actinomycetes which do not fix nitrogen also occur, there is much left to do, but work is still hampered by difficulties of isolation and the slow growth of these organisms.

Actinomycetaceae, Cellulomonadaceae, Microbacteriaceae and "*Micrococcaceae*"

Members of the families *Actinomycetaceae, Cellulomonadaceae, Microbacteriaceae* and "*Micrococcaceae*", along with *Arachnia, Brevibacterium* and *Dermatophilus*, were previously placed in the 'Actinobacteria'. This acted as a holding group for a heterogeneous assembly of genera which awaited the further work necessary before families could be proposed. As Table 14.1 shows it encompassed morphologically, physiologically and chemically quite different organisms. The groups of genera perceived showed spectra of morphological variation and imperfections in correlations of chemotaxonomic characters so that further genomic information was required. The *Cellulomonadaceae* is a good example, where members of the family are soil organisms which range from irregular rods and cocci (*Cellulomonas*) to branching hyphae fragmenting into motile rods (*Oerskovia*); these two genera have different cell wall chemistries, yet 16S rRNA sequencing indicates they share a line of descent (Fig. 14.2). The *Microbacteriaceae* contains organisms with the unusual B type of peptidoglycan (see Chapter 2) whose method of cross-linking is peculiar to this group and certain anaerobes. The two genera, *Curtobacterium* and *Clavibacter*, containing the majority of plant pathogens formerly in *Corynebacterium*, also belong to this family.

The family "*Micrococcaceae*" is chemotaxonomically and phylogenetically quite distinct (Table 14.1 & Fig. 14.2) and its members have coccoid morphology in common. *Micrococcus*, a genus of skin and environmental organisms, was for many years classified with the staphylococci, but cell wall analysis, its high mol% G+C and 16S rRNA sequencing indicate a relationship with the arthrobacters, representing, it has been suggested, a degenerate form stuck in the coccoid part of the life cycle. *Stomatococcus mucilaginosus*, an oral organism formerly in *Micrococcus*, and probably *Terrabacter tumescens* are further members of this family, but the affiliations of the chemotaxonomically similar *Renibacterium* and *Rothia* are unknown as yet.

The eponymous *Actinomyces* is the oldest genus of the large collection considered in this chapter. It was established in 1877 and the name was taken from the radial arrangement of filaments in granules taken from cattle with lumpy jaw. The genus is quite well circumscribed, but its internal taxonomy is complicated; the results of numerical taxonomies led to the suggestion that it might be split into three subgenera or genera, whereas whole-organism PAGE pattern clustering revealed two different groups that were further subdivisible into serovars. Conversely, DNA relatedness studies showed that some serovars warrant species status, but the lack of reliable phenotypic tests for differentiation (a problem throughout the genus) means they are left as genomic species. While 16S rRNA sequence comparisons confirm the distinctness of the species studied, they do not support splitting the genus into two or three genera.

Micromonosporaceae, Streptosporangiaceae and "*Thermomonosporaceae*"

The three informal and artificial aggregate groups, 'Actinoplanetes', 'Maduromycetes' and 'Thermomonosporas' were proposed in the wake of a series of unsatisfactory families, uncertainty about the true affiliations of their component genera, and in the light of numerical taxonomic and chemotaxonomic evidence. Actinoplanes and Maduromycetes formerly composed the family *Actinoplanaceae*, but peptidoglycan, whole-organism sugar and phospholipid analyses showed it to be divisible into two clusters which correlated with DNA relatedness; the *Thermomonospora* group comprised several morphologically diverse genera which have similar cell wall chemistries. Polyphasic approaches are now allowing taxonomic resolution within and between genera, *Actinomadura* being a good example.

The organism now called *A. madurae* spent 76 years in *Nocardia*, not without controversy, until the new genus *Actinomadura*, family *Thermoactinomycetaceae*, was proposed in 1970 in recognition of a special chemotaxonomic profile, particularly the presence of the new sugar madurose (Table 14.3). However, although new species were described and numerical taxonomy and menaquinone patterns supported generic status it was not accorded this in the 1974 edition of *Bergey's Manual*. With the emendation of its family to exclude non-endosporeformers the genus was accommodated in the new, and unsatisfactory as it proved, family *Thermomonosporaceae* until Maduromycetes was proposed as a temporary resting place for several genera of organisms with chemotype III cell walls and madurose, namely: *Actinomadura, "Excellospora", Microbispora, Microtetraspora, Planobispora, Planomonospora, Spirillospora* and *Streptosporangium*. The artificiality of the grouping is emphasized by the appearance of madurose in *Dermatophilus*, *Frankia*, and a streptomycete-like organism with a chemotype I cell wall and by DNA–rRNA hybridization and 16S rRNA cataloguing of four species. These studies showed *Actinomadura* to be a heterogeneous genus, with two species (including the type) related to *Thermomonospora* (Fig. 14.2).

Further major studies, in particular of phospholipids, fatty acids, menaquinones by HPLC, high-performance TLC and MS, DNA–rRNA hybridizations with *Nocardiopsis*, *Streptomyces* and *Streptosporangium*, DNA homology and numerical taxonomy, have confirmed the existence of two major clusters; members of the *A. madurae* group are of phospholipid type PI, have mainly 18-carbon 10-methyl branched fatty acids, and hexahydrogenated menaquinone with nine isoprene units as main components, whereas members of the *A. pusilla* group are of type PIV, have predominantly the 17-carbon fatty acid and the respective menaquinones are tetrahydrogenated (menaquinone data have not always been as clear-cut as this would suggest, however). *Microtetraspora* likewise appeared heterogeneous, most species being of phospholipid type PIV but one being of type PI.

Polyphasic studies led to revision of these genera: the *A. pusilla* group was transferred to *Microtetraspora* and the PI *Microtetraspora* was transferred to *Actinomadura*, the actinoplanetes *Micromonospora*, *Actinoplanes*, *Dactylosporangium* and *Pilimelia* were placed in an emended *Micromonosporaceae*, and the new family *Streptosporangiaceae* was proposed for *Microtetraspora*, *Microbispora*, *Planobispora*, *Planomonospora* and *Streptosporangium*. This still leaves *Actinomadura* without a family and the affiliations of the thermomonosporas remain unclear, but it is already apparent from 16S rRNA partial sequencing that *Actinomadura* and *Thermomonospora* are related at family level and that *Nocardiopsis* lies closer to the *Streptosporangiaceae*.

Important organisms

The actinomycetes are very widespread, perhaps ubiquitous, mainly saprophytic organisms, but some are parasites or commensals of humans and other animals or plants, and the order includes several important pathogens. The saprophytic species, chiefly of *Streptomyces* it seems, occur mainly in the soil, where they play important parts in organic decomposition and crumb structure and are responsible, in part, for its earthy smell. It is also believed that antibiotic production must have some effect on soil ecology, but reliable evidence for this is exceptionally difficult to obtain. Antibiotic production is certainly the best known property of these organisms and they are responsible for the vast majority of clinically and otherwise useful antibiotics; this is not only because they have been the most intensively studied and screened organisms. Although programmes are now experiencing diminishing returns, screening continues, as naturally occurring compounds often have the advantages of being less toxic, more biodegradable and cheaper and safer to produce than chemically synthesized drugs.

Of the actinomycetes the streptomycetes are pre-eminent as antibiotic producers, being the sources of such household names as the antibacterials streptomycin, chloramphenicol (subsequently produced by synthesis), neomycin, tetracyclines, erythromycin, vancomycin and cephamycin, the β-lactamase inhibitor clavulanic acid, and the antifungals nystatin and amphotericin B. There are also important contributions from *Micromonospora*, *Actinoplanes*, *Streptoverticillium* (now unified with *Streptomyces*), *Actinomadura* and *Nocardia* and several other genera. Most of the antibiotics are antibacterial and antifungal. Although antiviral, antitumour, insecticidal, herbicidal and antihelminthic drugs have been found, few of these have warranted commercial development. The search is continuing, in the hope of expansion of these areas and the discovery of drugs effective against protozoa and other parasites. About half of the world's antibiotic production is for non-clinical purposes, two of the more contentious uses being as growth promoters for livestock and as food preservatives. The streptomycetes are also the main producers of glucose isomerase for the food industry.

It is curious that members of the same order (and sometimes the same family or genus) that is such an important source of drugs also contains species pathogenic to humans and other animals. The most important group of pathogenic actinomycetes throughout history and at present is undoubtedly the mycobacteria. The first to be discovered was the cause of leprosy, *Mycobacterium leprae*, of which there are over 10 million cases worldwide. It has not yet been grown on artificial media but may be cultivated in footpads of mice and in nine-banded armadillos. The best known species are *M. tuberculosis*, the cause of tuberculosis in humans and some other animals, and *M. bovis*, which causes the bovine type but is also transmissible to man, particularly by milk, and is pathogenic for other animals. These three are responsible for the most mortality and morbidity, and tuberculosis cases are increasing (partly owing to HIV and antibiotic resistance), so that this disease is the world's leading cause of death from a single infectious agent, yet the first clinically useful streptomycete antibiotic, streptomycin, continues to be of value in the treatment of tuberculosis.

Other mycobacteria are of importance, not just as sources of confusion in sputum screening for tuberculosis diagnosis, but also as opportunists causing a range of infections, particularly pulmonary granulomatous lesions in immunocompromised persons (*M. avium*, especially important in AIDS patients, *M. intracellulare*, *M. kansasii* and *M. xenopi*), chronic cervical lymphadenitis in children (*M. scrofulaceum*), deep cutaneous ulceration (*M. ulcerans*) and superficial skin lesions (*M. marinum*). Animal diseases in addition to *M. bovis* infections include avian tuberculosis (*M. avium*), bovine farcy (*M. farcinogenes*, *M. senegalense*) and Johne's disease, a chronic diarrhoeal condition of cattle and sheep caused by *M. paratuberculosis*, which is of special medical interest because of its possible association with Crohn's disease, a chronic inflammatory bowel condition in humans.

Diseases caused by members of other groups include diphtheria (*Corynebacterium diphtheriae*), chronic purulent infections of farm animals

(*C. pseudotuberculosis*), bovine cystitis (*C. cystitidis*, *C. renale*), nosocomial infections with the antibiotic resistant species *C. jeikeium* (iatrogenic infections including septicaemia and endocarditis) and *C. urealyticum* (alkaline encrusted cystitis), chronic subcutaneous abscesses called actinomycotic mycetomas (*Actinomadura madurae*, *Actinomadura pelletieri*, *Nocardia brasiliensis*, and *Streptomyces somaliensis*), pulmonary nocardiosis (*N. asteroides*), actinomycosis (*Actinomyces israelii* and other species, and *Propionibacterium propionica*) which is characterized by chronic destructive abscesses of connective tissue discharging through non-healing sinuses, bovine actinomycosis or lumpy jaw (*Actinomyces bovis*), pyogenic disease of domestic animals (*Actinomyces pyogenes*), exudative dermatitis of domestic herbivores (*Dermatophilus congolensis*), foal bronchopneumonia (*Rhodococcus equi*), and chronic abscesses in salmonid fish (*Renibacterium salmoninarum*).

In addition, other members of several of the genera covered above, and species of *Arcanobacterium*, *Micrococcus*, *Nocardiopsis*, *Oerskovia*, *Rhodococcus* and *Rothia* may be opportunistic pathogens, and spore antigens of *Saccharomonospora* and *Saccharopolyspora* (*Faenia*) species growing in spontaneously heating vegetation cause extrinsic allergic alveolitis (or farmer's lung). Plant diseases include vascular wilts, stunts, gumming, and rots of food crops (*Clavibacter* species), vascular wilts and spots of food crops and ornamental plants (*Curtobacterium flaccumfaciens*), and potato scab (*Streptomyces scabies* and other species).

On the positive side, and in addition to organic decomposition by many actinomycetes and the production of antibiotics and enzymes by *Streptomyces* and other genera, *Frankia* species are of interest and importance as nitrogen fixers, *Rhodococcus* species are industrially important in the production of acrylic acid and acrylamide, and in steroid modification, and *Brevibacterium linens* and *Corynebacterium variabilis* have important roles in the ripening of soft cheeses such as Limburger.

Identification

Most actinomycete genera were originally defined on the basis of morphology, and although many definitions are now chemotaxonomic there have been relatively few radical changes at the genus level so that making some morphological observations on an unknown strain grown on a nutritionally poor medium remains the starting point for the identification of most isolates to genus level. However, even in the hands of very experienced persons, such an approach is not sufficiently reliable and chemotaxonomic data are valuable, particularly identification of the cell wall diamino acid and of the whole-organism sugar pattern by such simple methods as paper chromatography of whole-organism hydrolysates. Identification to species level is another matter. For genera containing only one or a few species a handful of confirmatory tests may be all that is required, but unfortunately these tend to be the less frequently encountered taxa, whereas species identification in such a large and continually expanding genus as *Streptomyces* can be daunting. Successful identification systems must be based on good classifications, and the many recent improvements in the taxonomy of the actinomycetes have made the task much easier.

Not surprisingly, much effort has been expended on improving the specificities, rapidities and simplicities of identification methods for organisms of medical, veterinary, industrial and agricultural importance. Rapid identification of mycobacteria is clearly desirable, but the slow growth rates of most of the medically important species means that traditional cultural and biochemical methods can take several weeks from primary isolation, which may itself be of low efficiency; some species are difficult to identify by these routine tests anyway. Methods have therefore been developed to identify from very small to tiny numbers of cells. Chemotaxonomic approaches include HPLC or pyrolysis GLC–MS of mycolic acid esters, which identify to genus level and can be of some value in identifying *M. leprae*, and capillary GLC and/or TLC of fatty acids, secondary alcohols and mycolate cleavage products, which together can differentiate most of the species. Immunological methods include immunodiffusion and ELISA, but like the chemotaxonomic methods 2–3 weeks are usually needed to obtain sufficient material (perhaps 10^6–10^7 cells) for analysis.

Genomic methods, however, can offer immediate identification (i.e. a few hours to a few days) by making cultivation unnecessary. Sensitivities of down to one cell, so permitting direct identification in clinical specimens, have been claimed for nucleic acid probes and many methods and applications, with the mycobacteria prominent, have been published. Target sequences include those of specific antigens or of highly variable and so species-specific regions of 16S rRNA, and the most rapid and sensitive rely on amplification of DNA using the polymerase chain reaction. These methods have also been applied to related groups; examples are capillary GLC–MS of lipids and mycolates for *Corynebacterium urealyticum*, and immunoblotting and DNA probes for nocardiosis diagnosis, which is slow and difficult by clinical signs, radiology and histology.

Identification of *Streptomyces* species relied, until recently, on morphology and pigmentation, but a method that is so subjective and difficult to standardize must often be unsuccessful. It is the very importance of these organisms, resulting in the frequent description of new species or so-called species (often based on a single strain, or very few isolates from a single source), that makes a satisfactory classification (and therefore an effective identification scheme) so hard to achieve and yet so desirable. Modern approaches to the problem include numerical taxonomy, nucleic acid hybridization, ELISA of antigens in whole-organism lysates, rapid biochemical assay, electrophoresis of whole-organism extracts and two-dimensional PAGE of ribosomal proteins.

The most comprehensive studies, and the only ones that can be said to have generated an identification scheme, are those of Williams' and Kroppenstedt's groups, which have concentrated on numerical taxonomies of the well-established species. The former revealed major (≥4 member), minor (2–3 member) and single member clusters, which were then separated into three respective matrices, two probabilistic (% positives) for computer identification and one as a diagnostic table (+/– results). Characters, which included morphology, pigmentation, antibiosis, antibiotic sensitivity, growth tolerances and nutritional requirements, were selected by computer for their separation values and diagnostic power for individual clusters, to reduce them to a manageable number. However, the diversity of the streptomycetes is reflected by the need for three matrices rather than just one to cover the 78 taxa, and the large numbers of tests (50, 39 and 37) that they contain. Likewise, Kroppenstedt's group used 50 miniaturized utilization and hydrolysis tests in microtitration plates to identify 52 well-represented taxa. In both cases the requirements for many special media or substrates would make occasional identification in a non-specialist laboratory an expensive business. None the less, the schemes were 77–78% successful in trials and the databases can be supplemented as new strains and species are characterized.

Further reading

Books of general interest and relevant to more than one chapter

Balows, A., Trüper, H.G., Dworkin, M., Harder, W. & Schleifer, K.-H. (1991) *The Prokaryotes*, 2nd edn. Springer-Verlag, New York.

Holt, J.G., Krieg, N.R., Sneath, P.H.A., Staley, J.T. & Williams, S.T. *Bergey's Manual of Systematic Bacteriology*. Williams & Wilkins, Baltimore.

Volume 1 (1984) Spirochaetes, Gram-negative bacteria, rickettsiae, chlamydiae and mollicutes.

Volume 2 (1986) Gram-positive bacteria.

Volume 3 (1989) Autotrophs, budding bacteria, gliding bacteria, archaebacteria.

Volume 4 (1989) Actinomycetes.

Journals

Original articles on systematics appear in:

International Journal of Systematic Bacteriology
Systematic and Applied Microbiology

Articles on classification and identification regularly appear in:

Antonie van Leeuwenhoek
Applied and Environmental Microbiology
Archives of Microbiology
Canadian Journal of Microbiology
Journal of Applied Bacteriology
Journal of Bacteriology
Journal of Clinical Microbiology
Journal of General Microbiology
Journal of General and Applied Microbiology
Letters in Applied Microbiology

Chapter 1: Introduction

Bulloch, W. (1960) *History of Bacteriology*. Oxford University Press, London.

Cowan, S.T. & Hill, L.R. (1978) *A Dictionary of Microbial Taxonomy*. Cambridge University Press, Cambridge.

Jeffrey, C. (1989) *Biological Nomenclature*, 3rd edn. Edward Arnold, London

Moore, W.E.C. & Moore, L.V.H. (1989) *Index of the Bacterial and Yeast Nomenclatural Changes*. American Society for Microbiology, Washington, D.C.

Skerman, V.B.D., McGowan, V. & Sneath, P.H.A. (1989) *Approved Lists of Bacterial Names*. American Society for Microbiology, Washington, D.C.

Sneath, P.H.A. (1957) The application of computers to taxonomy. *Journal of General Microbiology* **17**, 201–226.

Sneath, P.H.A. (ed.) (1992) *International Code of Nomenclature of Bacteria 1990 Revision*. American Society for Microbiology, Washington, D.C.

Sneath, P.H.A. & Sokal, R.R. (1973) *Numerical Taxonomy*. W. H. Freeman, San Francisco.

Chapters 2 & 3: Phenotypic and genotypic characters

Colwell, R.R. & Grigorova, R. (eds) (1987) *Methods in Microbiology, Volume 19 – Current Methods for Classification and Identification of Microorganisms.* Academic Press, London.

Dykhuizen, D.E. & Green, L. (1991) Recombination in *Escherichia coli* and the definition of biological species. *Journal of Bacteriology* **173**, 7257–7268.

Ezaki, T., Hashimoto, Y. & Yabuuchi, E. (1989) Fluorometric DNA–DNA hybridization in microdilution wells as an alternative to membrane filter hybridization in which radioisotopes are used to determine genetic relatedness among bacterial strains. *International Journal of Systematic Bacteriology* **39**, 224–229.

Goodfellow, M. & Minnikin, D.E. (eds) (1985) *Chemical Methods in Bacterial Systematics*. Academic Press, London.

Goodfellow, M. & O'Donnell, A.G. (eds) (1993) *Handbook of New Bacterial Systematics*. Academic Press, London.

Magee, J.T., Hindmarch, J.M., Bennett, K.W., Duerden, B.I. & Aries, R.E. (1989) A pyrolysis mass spectrometry study of fusobacteria. *Journal of Medical Microbiology* **28**, 227–236.

Mesbah, M., Premachandran, U. & Whitman, W. (1989) Precise measurement of the G+C content of deoxyribonucleic acid by high-performance liquid chromatography. *International Journal of Systematic Bacteriology* **39**, 159–167.

Monoharan, R., Ghiamati, E., Daltiero, R.A., Britton, K.A., Nelson, W.H. & Sperry, J.F. (1990) UV resonance Raman spectra of bacteria, bacterial spores, protoplasts and calcium dipicolinate. *Journal of Microbiological Methods* **11**, 1–15.

Scherer, P. & Kneifel, H. (1983) Distribution of polyamines in methanogenic bacteria. *Journal of Bacteriology* **154**, 1315–1322.

Sneath, P.H.A. & Johnson, R. (1972) The influence on numerical taxonomic similarities of errors in microbiological tests. *Journal of General Microbiology* **72**, 377–392.

Stackebrandt, E. & Goodfellow, M. (eds) (1991) *Nucleic Acid Techniques in Bacterial Systematics*. John Wiley & Sons, Chichester.

Weisburg, W.G., Barns, S.M., Pelletier, D.A. & Lane D.J. (1991) 16S ribosomal DNA amplification for phylogenetic study. *Journal of Bacteriology* **173**, 697–703.

Chapter 4: Similarity and arrangement

Abbott, L.A., Bisby, F.A. & Rogers, D.J. (1985) *Taxonomic Analysis in Biology*. Columbia University Press, New York.

Bryant, T.N. & Wimpenny, J.W.T. (eds) (1989) *Computers in Microbiology*. IRL Press, Oxford.

Goodfellow, M., Jones, D. & Priest, F.G. (eds) (1985) *Computer-Assisted Bacterial Systematics*. Academic Press, London.

Sneath, P.H.A. (1989) Analysis and interpretation of sequence data for bacterial systematics: the view of a numerical taxonomist. *Systematic and Applied Microbiology* **12**, 15–31.

Sneath, P.H.A. & Johnson, R. (1972) The influence on numerical taxonomic similarities of errors in microbiological tests. *Journal of General Microbiology* **72**, 377–392.

Sneath, P.H.A. & Sokal, R.R. (1973) *Numerical Taxonomy.* W.H. Freeman, San Francisco.

Chapter 5: Identification

Barrow, G.I. & Feltham, R.K.A. (eds) (1993) *Cowan and Steel's Manual for the Identification of Medical Bacteria,* 3rd edn. Cambridge University Press, Cambridge.

Board, R.G., Jones, D. & Skinner, F.E. (eds) (1992) *Identification Methods in Applied and Environmental Microbiology.* Blackwell Scientific Publications, Oxford.

Collins, C.H. & Grange, J.M. (eds) (1985) *Isolation and Identification of Microorganisms of Medical and Veterinary Importance.* Academic Press, London.

DeLong, E.F., Wickham, G.S. & Pace, N.R. (1989) Phylogenetic stains: ribosomal RNA-based probes for the identification of single cells. *Science* **243**, 1361–1363.

Freney, J., Herve, C., Desmonceaux, M., Allard, F., Boeufgras, J.M., Monget, D. & Fleurette, J. (1991) Description and evaluation of the semiautomated 4-hour ATB 32E method for identification of members of the family *Enterobacteriaceae. Journal of Clinical Microbiology* **29,** 138–141.

Helm, D., Labischinski, H., Schallehn, G. & Naumann, D. (1991) Classification and identification of bacteria by Fourier-transform infrared spectroscopy. *Journal of General Microbiology* **137,** 69–79.

Kämpfer, P. & Altwegg, M. (1992) Numerical classification and identification of *Aeromonas* genospecies. *Journal of Applied Bacteriology* **72,** 341–351.

Rowe, M.T. & Finn, B. (1991) A study of *Pseudomonas fluorescens* biovars using the Automated Microbiology Identification System (AMBIS). *Letters in Applied Microbiology* **13**, 238–242.

Chapter 6: Evolution and the archaea

Auer, J., Spicker, G., Mayerhofer, L., Pühler, G. & Böch, A. (1991) Organisation and nucleotide sequence of a gene cluster comprising the translation elongation factor 1α from *Sulfolobus acidocaldarius. Systematic and Applied Microbiology* **14**, 14–22.

Iwabe, N., Kuma, K.-I., Hasegawa, M., Osawa, S. & Miyata, T. (1989) Evolutionary relationship of archaebacteria, eubacteria, and eukaryotes inferred from phylogenetic trees of duplicated genes. *Proceedings of the National Academy of Sciences of the USA* **86**, 9355–9359.

Liu, Y., Boone, D.R. & Choy, C. (1990) *Methanohalophilus oregonense* sp. nov., a methylotrophic methanogen from an alkaline, saline aquifer. *International Journal of Systematic Bacteriology* **40**, 111–116.

Macario, A.J.L., Dugan, C. B. & Conway de Macario, E. (1987) Antigenic mosaic of *Methanogenium* spp.: analysis with poly- and monoclonal-antibody probes. *Journal of Bacteriology* **169**, 666–669.

Pierson, B.K. & Thornbeer, J.P. (1983) Isolation and special characterization of photochemical reaction centers from the thermophilic green bacterium *Chloroflexus aurantiacus* strain J-10-f1. *Proceedings of the National Academy of Sciences of the USA* **80**, 80–84.

Woese, C.R. (1987) Bacterial evolution. *Microbiological Reviews* **51**, 221–271.
Woese, C.R., Kandler O. & Wheelis, M.L. (1990) Towards a natural system of organisms: proposal for the domains Archaea, Bacteria, and Eucarya. *Proceedings of the National Academy of Sciences of the USA* **87**, 4576–4579.
Zillig, W., Holz, I. & Wunderl, S. (1991) *Hyperthermus butylicus*, gen. nov., sp. nov. a hyperthermophilic, anaerobic, peptide fermenting, facultatively H_2S-generating archaebacterium. *International Journal of Systematic Bacteriology* **41,** 169–170.

Comprehensive coverage of the following groups will be found in *Bergey's Manual* and *The Prokaryotes*; useful background reading, important recent taxonomic studies, and papers covering recent developments and illustrating the application of particular taxonomic methods are cited below.

Chapter 7: The spirochaetes

Baranton, G., Postic, D., Saint Girons, I., Boerlin, P., Piffaretti, J.-C., Assous, M. & Grimont, P.A.D. (1992) Delineation of *Borrelia burgdorferi* sensu stricto, *Borrelia garinii* sp. nov., and group VS461 associated with Lyme borreliosis. *International Journal of Systematic Bacteriology* **42**, 378–383.
Cacciapuoti, B., Ciceroni, L. & Attard-Barbini, D. (1991) Fatty acid profiles, a chemotaxonomic key for the classification of strains of the family *Leptospiraceae. International Journal of Systematic Bacteriology* **41**, 295–300.
Hookey, J.V., Waitkins, S.A. & Jackman, P.J.H. (1985) Numerical analysis of leptospira DNA-restriction endonuclease patterns. *FEMS Microbiology Letters* **29**, 185–188.
Paster, B.J., Dewhirst, F.E., Weisburg, W.G., Tordoff, L.A., Fraser, G.J., Hespell, R.B., Stanton, T.B., Zablen, L., Mandelco, L. & Woese, C.R. (1991) Phylogenetic analysis of the spirochaetes. *Journal of Bacteriology* **173**, 6101–6109.
Ramadass, P., Jarvis, B.D.W., Corner, R.J., Cinco, M. & Marshall, R.B. (1990) DNA relatedness among strains of *Leptospira biflexa. International Journal of Systematic Bacteriology* **40**, 231–235.
Ramadass, P., Jarvis, B.D.W., Corner, R.J., Penny, D. & Marshall, R.B. (1992) Genetic characterization of pathogenic *Leptospira* species by DNA hybridization. *International Journal of Systematic Bacteriology* **42,** 215–219.
Stanton, T.B., Jensen, N.S., Casey, T.A., Tordoff, L.A., Dewhirst, F.E. & Paster, B.J. (1991) Reclassification of *Treponema hyodysenteriae* and *Treponema innocens* in a new genus, *Serpula* gen. nov., as *Serpula hyodysenteriae* comb. nov. and *Serpula innocens* comb. nov. *International Journal of Systematic Bacteriology* **41**, 50–58. (*Serpula* was later changed to *Serpulina*).
Yasuda, P.H., Steigerwalt, A.G., Sulzer, K.R., Kaufmann, A.F., Rogers, F. & Brenner, D.J. (1987) Deoxyribonucleic acid relatedness between serogroups and serovars in the family *Leptospiraceae* with proposals for seven new *Leptospira* species. *International Journal of Systematic Bacteriology* **37**, 407–415.

Chapter 8: Helical and curved bacteria

Lane, D.J., Harrison, A.P., Stahl, D., Pace, B., Giovannoni, S.J., Olsen, G.J. & Pace, N.R. (1992) Evolutionary relationships among sulfur- and iron-oxidizing bacteria. *Journal of Bacteriology* **174**, 269–278.

Lee, A., Phillips, M.W., O'Rourke, J.L., Paster, B.J., Dewhirst, F.E., Fraser, G.J., Fox, J.G., Sly, L.I., Romaniuk, P.J., Trust, T.J. & Kouprach, S. (1992) *Helicobacter muridarum* sp. nov., a microaerophilic helical bacterium with a novel ultrastructure isolated from the intestinal mucosa of rodents. *International Journal of Systematic Bacteriology* **42**, 27–36.

McNulty, C.A.M., Dent, J.C., Curry, A., Uff, J.S., Ford, G.A., Gear, M.W.L. & Wilkinson, S.P. (1989) New spiral bacterium in gastric mucosa. *Journal of Clinical Pathology* **42**, 585–591.

Pot, B., Willems, A., Gillis, M. & De Ley, J. (1992) Intra- and intergeneric relationships of the genus *Aquaspirillum*: *Prolinoborus*, a new genus for *Aquaspirillum fasciculus*, with the species *Prolinoborus fasciculus* comb. nov. *International Journal of Systematic Bacteriology* **42**, 44–57.

Stanley, J., Burnens, A.P., Linton, D., On, S.W.L., Costas, M. & Owen, R.J. (1991) *Campylobacter helveticus* sp. nov., a new thermophilic species from domestic animals; characterization and cloning of a species-specific DNA probe. *Journal of General Microbiology* **138**, 2293–2303.

Thompson, L.M., Smibert, R.M., Johnson, J.L. & Krieg, N.R. (1988) Phylogenetic study of the genus *Campylobacter. International Journal of Systematic Bacteriology* **38**, 190–200.

Vandamme, P. & De Ley, J. (1991) Proposal for a new family, *Campylobacteraceae. International Journal of Systematic Bacteriology* **41**, 451–455.

Vandamme, P., Falsen, E., Rossau, R., Hoste, B., Segers, P., Tytgat, R. & De Ley, J. (1991) Revision of *Campylobacter*, *Helicobacter*, and *Wolinella* taxonomy: emendation of generic descriptions and proposal of *Arcobacter* gen. nov. *International Journal of Systematic Bacteriology* **41**, 88–103.

Chapter 9: Gram-negative aerobic bacteria

Allardet-Servent, A., Bourg, G., Ramuz, M., Pages, M., Bellis, M. & Roizes, G. (1988) DNA polymorphism in strains of the genus *Brucella. Journal of Bacteriology* **170**, 4603–4607.

Bøvre, K. (1970) Pulse-RNA–DNA hybridization between rod-shaped and coccal species of the *Moraxella–Neisseria* groups. *Acta Pathologica Microbiologica et Immunologica Scandinavica* **78B**, 565–574.

Brindle, R.J., Bryant, T.N. & Draper, P.W. (1989) Taxonomic investigation of *Legionella pneumophila* using monoclonal antibodies. *Journal of Clinical Microbiology* **27**, 536–539.

Byng, G.S., Whitaker, R.J., Gherna, R.L. & Jensen, R.A. (1980) Variable enzymological patterning in tyrosine biosynthesis as a means of determining natural relatedness among the *Pseudomonadaceae. Journal of Bacteriology* **144**, 247–257.

Dennis, P.J., Brenner, D.J., Thacker, W.L., Wait, R., Vesey, G., Steigerwalt, A.G. & Benson, R.F. (1993) Five new *Legionella* species isolated from water. *International Journal of Systematic Bacteriology* **43,** 329–337.

Dudman, W.F. & Belbin, L. (1988) Numerical taxonomic analysis of some strains of *Rhizobium* that uses a qualitative coding of immunodiffusion reactions. *Applied and Environmental Microbiology* **54**, 1825–1830.

Fry, N.K., Warwick, S., Saunders, N.A. & Embley, T.M. (1991) The use of 16S ribosomal RNA analyses to investigate the phylogeny of the family *Legionellaceae. Journal of General Microbiology* **137,** 1215–1222.

Lambert, M.A. & Moss C.W. (1989) Cellular fatty acid compositions and isoprenoid quinone contents of 23 *Legionella* species. *Journal of Clinical Microbiology* **27**, 465–473.

Moreno, E., Stackebrandt, E., Dorsch, M., Wolters, J., Busch, M. & Mayer, H. (1990) *Brucella abortus* 16S rRNA and lipid A reveal a phylogenetic relationship with members of the alpha-2 subdivision of the class *Proteobacteria. Journal of Bacteriology* **172**, 3569–3576.

Sneath, P.H.A., Stevens, M. & Sackin, M.J. (1981) Numerical taxonomy of *Pseudomonas* based on published records of substrate utilization. *Antonie van Leeuwenhoek* **47**, 423–448.

Thompson, J.P. & Skerman, V.B.D. (eds) (1979) *Azotobacteraceae: the Taxonomy and Ecology of the Aerobic Nitrogen Fixing Bacteria*. Academic Press, London.

Vaneechoutte, M., Rossau, R., De Vos, P., Gillis, M., Janssens, D., Paepe, N., De Rouck, A., Fiers, T., Claeys, G. & Kersters, K. (1992) Rapid identification of bacteria of the *Comamonadaceae* with amplified ribosomal DNA-restriction analysis (ARDRA). *FEMS Microbiology Letters* **93**, 227–234.

Willems, A., Falsen, E., Pot, B., Jantzen, E., Hoste, B., Vandamme, P., Gillis, M., Kersters, K. & De Ley, J. (1990) *Acidovorax*, a new genus for *Pseudomonas facilis, Pseudomonas delafieldii*, E. Falsen (EF) group 16, and several clinical isolates, with the species *Acidovorax facilis* comb. nov., *Acidovorax delafieldii* comb. nov., and *Acidovorax temperans* sp. nov. *International Journal of Systematic Bacteriology* **40**, 384–398.

Chapter 10: Gram-negative, facultatively and strictly anaerobic bacteria

Brayton, P.R., Bode, R.B., Colwell, R.R., MacDonell, M.T., Hall, H.L., Grimes, D.J., West, P.A. & Bryant, T.N. (1986) *Vibrio cincinnatiensis* sp. nov., a new human pathogen. *Journal of Clinical Microbiology* **23**, 104–108.

Buchanan, R.E., Holt, J.G. & Lessel, E.F. (1966) *Index Bergeyana.* Williams & Wilkins, Baltimore.

De Ley, J., Mannheim, W., Mutters, R., Piechulla, K., Tytgat, R., Segers, P., Bisgaard, M., Frederiksen, W., Hinz, K.-H. & Vanhoucke, M. (1990) Inter- and intrafamilial similarities of rRNA cistrons of the *Pasteurellaceae. International Journal of Systematic Bacteriology* **40**, 126–137.

Dewhirst, F.E., Paster, B.J., Olsen, I. & Fraser, G.J. (1992) Phylogeny of 54 representative strains of species in the family *Pasteurellaceae* as determined by comparison of 16S rRNA sequences. *Journal of Bacteriology* **174**, 2002–2013.

Farmer, J.J., Davis, B.R., Hickman-Brenner, F.W., McWhorter, A., Huntley-Carter, G.P., Asbury, M.A., Riddle, C., Wathen-Grady, H.G., Elias, C., Fanning, G.R., Steigerwalt, A.G., O'Hara, C.M., Morris, G.K., Smith, P.B. & Brenner, D.J. (1985) Biochemical identification of new species and biogroups of *Enterobacteriaceae* isolated from clinical specimens. *Journal of Clinical Microbiology* **21**, 46–76.

Gharbia, S.E. & Shah, H.N. (1990) Identification of *Fusobacterium* species by the electrophoretic migration of glutamate dehydrogenase and 2 oxoglutarate reductase in relation to their DNA base composition and peptidoglycan dibasic amino acids. *Journal of Medical Microbiology* **33**, 183–188.

Gibbons, N.E., Pattee, K.B. & Holt, J.G. (1981) *Supplement to Index Bergeyana.*

Williams & Wilkins, Baltimore.

Lawrence, J.G., Ochman, H. & Hartl, D.L. (1991) Molecular and evolutionary relationships among enteric bacteria. *Journal of General Microbiology* **137**, 1911–1921.

Martinez-Murcia, A.J., Benlloch, S. & Collins, M.D. (1992) Phylogenetic interrelationships of members of the genera *Aeromonas* and *Plesiomonas* as determined by 16S ribosomal DNA sequencing: lack of congruence with results of DNA–DNA hybridizations. *International Journal of Systematic Bacteriology* **42**, 412–421.

Mergaert, J., Verdonck, L., Kersters, K., Swings, J., Boeufgras, J.-M. & De Ley, J. (1984) Numerical taxonomy of *Erwinia* species using API systems. *Journal of General Microbiology* **130**, 1893–1910.

Myhrvold, V., Brondz, I. & Olsen, I. (1992) Application of multivariate analyses of enzymic data to classification of members of the *Actinobacillus–Haemophilus–Pasteurella* group. *International Journal of Systematic Bacteriology* **42**, 12–18.

Picard-Pasquier, N., Picard, B., Heeralal, S., Krishnamoorthy, R. & Goullet, P. (1990) Correlation between ribosomal DNA polymorphism and electrophoretic enzyme polymorphism in *Yersinia*. *Journal of General Microbiology* **136**, 1655–1666.

Suhail, A., Weisburg, W.G. & Jensen, R.A. (1990) Evolution of aromatic amino acid biosynthesis and application to the fine-tuned phylogenetic positioning of enteric bacteria. *Journal of Bacteriology* **172**, 1051–1061.

Chapter 11: The Gram-positive rods

Ash, C., Farrow, J.A.E., Wallbanks, S. & Collins, M.D. (1991) Phylogenetic heterogeneity of the genus *Bacillus* revealed by comparative analysis of small-subunit-ribosomal RNA sequences. *Letters in Applied Microbiology* **13**, 202–206.

Berkeley, R.C.W., Logan, N.A., Shute, L.A. & Capey, A.G. (1984) Identification of *Bacillus* species. In Bergan, T. (ed) *Methods in Microbiology*, Volume 16, pp. 291–328. Academic Press, London.

Boerlin, P., Rocourt, J. & Piffaretti, J.-C. (1991) Taxonomy of the genus *Listeria* by using multilocus enzyme electrophoresis. *International Journal of Systematic Bacteriology* **41**, 59–64.

Cato, E.P. & Stackebrandt, E. (1989) Taxonomy and phylogeny. In Minton, N.P. & Clarke, D.J. (eds) *Clostridia*, pp. 1–26. Plenum Press, New York.

Gurtler, V., Wilson, V.A. & Mayall, B.D. (1991) Classification of medically important clostridia using restriction endonuclease site differences of PCR-amplified 16S rDNA. *Journal of General Microbiology* **137**, 2673–2679.

Johnson, J.L. & Francis, B.S. (1975) Taxonomy of the clostridia: ribosomal ribonucleic acid homologies among the species. *Journal of General Microbiology* **88**, 229–244.

Lee, W.H. & Riemann, H. (1970) Correlation of toxic and non-toxic strains of *Clostridium botulinum* by DNA composition and homology. *Journal of General Microbiology* **60**, 117–123.

Priest, F.G. & Barbour, E.A. (1985) Numerical taxonomy of lactic acid bacteria and some related taxa. In Goodfellow, M., Jones, D. & Priest, F.G. (eds) *Computer-Assisted Bacterial Systematics*, pp. 137–163. Academic Press, London.

Ståhl, M., Molin, G., Persson, A., Ahrné, S. & Ståhl, S. (1990) Restriction endonuclease patterns and multivariate analysis as a classification tool for *Lactobacillus* spp. *International Journal of Systematic Bacteriology* **40**, 189–193.

Wisotzkey, J.D., Jurtshuk, P., Fox, G.E., Deinhard, G. & Poralla, K. (1992) Comparative sequence analyses on the 16S rRNA (rDNA) of *Bacillus acidocaldarius, Bacillus acidoterrestris,* and *Bacillus cycloheptanicus* and proposal for creation of a new genus, *Alicyclobacillus* gen. nov. *International Journal of Systematic Bacteriology* **42**, 263–269.

Chapter 12: The Gram-positive cocci

Bentley, R.W., Leigh, J.A. & Collins, M.D. (1991) Intrageneric structure of *Streptococus* based on comparative analysis of small-subunit rRNA sequences. *International Journal of Systematic Bacteriology* **41**, 487–494.

Dicks, L.M.T., van Vuuren, H.J.J. & Dellaglio, F. (1990) Taxonomy of *Leuconostoc* species, particularly *Leuconostoc oenos,* as revealed by numerical analysis of total soluble cell protein patterns, DNA base compositions, and DNA–DNA hybridizations. *International Journal of Systematic Bacteriology* **40**, 83–91.

French, G.L., Talsania, H., Charlton, J.R.H. & Phillips, I. (1989) A physiological classification of viridans streptococci by use of the API-20STREP system. *Journal of Medical Microbiology* **28**, 275–286.

Gilmour, M.N., Whittam, T.S., Kilian, M. & Selander, R.K. (1987) Genetic relationships among the oral streptococci. *Journal of Bacteriology* **169**, 5247–5257.

Hájek, V., Ludwig, W., Schleifer, K.H., Springer, N., Zitzelsberger, W., Kroppenstedt, R.M. & Kocur, M. (1992) *Staphylococcus muscae,* a new species isolated from flies. *International Journal of Systematic Bacteriology* **42,** 97–101.

Jones, D., Board, R.G. & Sussman, M. (eds) (1990) *Staphylococci.* Society for Applied Bacteriology Symposium Series No. 19. *Journal of Applied Bacteriology* **69**. Supplement.

Oyaizu, H., Stackebrandt, E., Schleifer, K.H., Ludwig, W., Pohla, H., Ito, H., Hirata, A., Oyaizu, Y. & Komagata, K. (1987) A radiation-resistant rod-shaped bacterium, *Deinobacter grandis* gen. nov., sp. nov., with peptidoglycan containing ornithine. *International Journal of Systematic Bacteriology* **37**, 62–67.

Pompei, R., Thaller, M.C., Pittaluga, F., Flore, O. & Satta, G. (1992) Analysis of bacteriolytic activity patterns, a novel approach to the taxonomy of the enterococci. *International Journal of Systematic Bacteriology* **42**, 37–43.

Schleifer, K.H. & Kilpper-Bälz, R. (1987) Molecular and chemotaxonomic approaches to the classification of streptococci, enterococci and lactococci: a review. *Systematic and Applied Microbiology* **10**, 1–9.

Weisburg, W.G., Giovannoni, S.J. & Woese, C.R. (1989) The *Deinococcus–Thermus* phylum and the effect of rRNA composition on phylogenetic tree construction. *Systematic and Applied Microbiology* **11**, 128–134.

Chapter 13: The mollicutes

Barile, M.F., Razin, S., Tully, J.G. & Whitcomb, R.F. (1979 on) *The Mycoplasmas* Volumes I–V. Academic Press, San Diego & London.

Costas, M., Leach, R.H. & Mitchelmore, D.L. (1987) Numerical analysis of PAGE protein patterns and the taxonomic relationships within the *Mycoplasma mycoides* cluster. *Journal of General Microbiology* **133**, 3319–3329.

Grau, O., Laigret, F., Carle, P., Tully, J.G., Rose, D.L. & Bové, J.M. (1991) Identification of a plant-derived mollicute as a strain of an avian pathogen, *Mycoplasma iowae*, and its implications for mollicute taxonomy. *International Journal of Systematic Bacteriology* **41**, 473–478.

Kuske, C.R. & Kirkpatrick, B.C. (1992) Phylogenetic relationships between the western aster yellows mycoplasma-like organism and other prokaryotes established by 16S rRNA gene sequence. *International Journal of Systematic Bacteriology* **42**, 226–233.

Kuske, C.R., Kirkpatrick, B.C. & Seemüller, E. (1991) Differentiation of virescence MLOs using western aster yellows mycoplasma-like organism chromosomal DNA probes and restriction fragment length polymorphism analysis. *Journal of General Microbiology* **137**, 153–159.

Lim, P.O. & Sears, B.B. (1989) 16S rRNA sequence indicates that plant-pathogenic mycoplasma-like organisms are evolutionarily distinct from animal mycoplasmas. *Journal of Bacteriology* **171**, 5901–5906.

Robinson, I.M. & Freundt, E.A. (1987) Proposal for an amended classification of anaerobic mollicutes. *International Journal of Systematic Bacteriology* **37**, 78–81.

Taylor, T.K., Bashiruddin, J.B. & Gould, A.R. (1992) Relationships between members of the *Mycoplasma mycoides* cluster as shown by DNA probes and sequence analysis. *International Journal of Systematic Bacteriology* **42**, 593–601.

Tully, J.G., Bové, J.M., Laigret, F. & Whitcomb, R.F. (1993) Revised taxonomy of the class *Mollicutes*: proposed elevation of a monophyletic cluster of arthropod-associated mollicutes to ordinal rank (*Entomoplasmatales* ord. nov.) with provision for familial rank to separate species with nonhelical morphology (*Entomoplasmataceae* fam. nov.) from helical species (*Spiroplasmataceae* fam. nov.) and emended descriptions of the order *Mycoplasmatales*, family *Mycoplasmataceae*. *International Journal of Systematic Bacteriology* **43**, 378–385.

Weisburg, W.G., Tully, J.G., Rose, D.L., Petzel, J.P., Oyaizu, H., Yang, D., Mandelco, L., Sechrest, J., Lawrence, T.G., Van Etten, J., Maniloff, J. & Woese, C.R. (1989) A phylogenetic analysis of the mycoplasmas: basis for their classification. *Journal of Bacteriology* **171**, 6455–6467.

Chapter 14: The actinomycetes

Böddinghaus, B., Rogall, T., Flohr, T., Blöcker, H. & Böttger, E.C. (1990) Detection and identification of mycobacteria by amplification of rRNA. *Journal of Clinical Microbiology* **28**, 1751–1759.

Collins, M.D., Smida, J., Dorsch, M. & Stackebrandt, E. (1988) *Tsukamurella* gen. nov. harboring *Corynebacterium paurometabolum* and *Rhodococcus aurantiacus*. *International Journal of Systematic Bacteriology* **38**, 385–391.

Embley, T.M., Smida, J. & Stackebrandt, E. (1988) The phylogeny of mycolateless wall chemotype IV actinomycetes and description of *Pseudonocardiaceae* fam. nov. *Systematic and Applied Microbiology* **11**, 44–52.

Goodfellow, M., Stanton, L.J., Simpson, K.E. & Minnikin, D.E. (1990) Numerical and chemical classification of *Actinoplanes* and some related actinomycetes. *Journal of General Microbiology* **136**, 19–36.

Langham, C.D., Williams, S.T., Sneath, P.H.A. & Mortimer, A.M. (1989) New probability matrices for identification of *Streptomyces*. *Journal of General Microbiology* **135**, 121–133.

Lévy-Frébault, V.V., Thorel, M.-F., Varnerot, A. & Gicquel, B. (1989) DNA polymorphism in *Mycobacterium paratuberculosis*, 'wood pigeon mycobacteria', and related mycobacteria analyzed by field inversion gel electrophoresis. *Journal of Clinical Microbiology* **27**, 2823–2826.

Luquin, M., Ausina, V., López Calahorra, F., Belda, F., García Barceló, M., Celma, C. & Prats, G. (1991) Evaluation of practical chromatographic procedures for identification of clinical isolates of mycobacteria. *Journal of Clinical Microbiology* **29**, 120–130.

Pitulle, C., Dorsch, M., Kazda, J., Wolters, J. & Stackebrandt, E. (1992) Phylogeny of rapidly growing members of the genus *Mycobacterium*. *International Journal of Systematic Bacteriology* **42,** 337–343.

St-Laurent, L., Bousquet, J., Simon, L. & Lalonde, M. (1987) Separation of various *Frankia* strains in the *Alnus* and *Elaeagnus* host specificity groups using sugar analysis. *Canadian Journal of Microbiology* **33**, 764–772.

Wellington, E.M.H., Stackebrandt, E., Sanders, D., Wolstrup, J. & Jorgensen, N.O.G. (1992) Taxonomic status of *Kitasatosporia*, and proposed unification with *Streptomyces* on the basis of phenotypic and 16S rRNA analysis and emendation of *Streptomyces* Waksman and Henrici 1943, 339[AL]. *International Journal of Systematic Bacteriology* **42**, 156–160.

Organism index

This index includes bacteria and groups of bacteria under their scientific names only.
For other references please use the Subject index.
Page numbers in *italics* indicate tables or figures.

Subject index

This index includes references to bacteria under their common names. The Organisms index includes bacteria under their scientific names only.
Page numbers in **bold** indicate definitions or explanations of terms or concepts; those in *italics* indicate figures or tables.